MW00720015

Experiences of Gold Hunters in Alaska

BY

CHARLES A. MARGESON

Preface by Jim and Nancy Lethcoe

PUBLISHED BY THE AUTHOR

1899

Reprinted by Prince William Sound Books
Valdez, Alaska
1997

If you have enjoyed reading this book, you may want to read other books on the history of this area published by Prince William Sound Books:

Jim and Nancy Lethcoe, *A History of Prince William Sound.* 1994. 160 pp. 15 maps and illustrations, 50 photos, ISBN 1-877900-04-4. $14.95.

Jim and Nancy Lethcoe, *Valdez Gold Rush Trails of 1898-99.* 1996. 144 pp, 8 maps, 47 photos. ISBN 1-877900-05-2, $14.95.

Addison M. Powell, *Trailing and Camping in Alaska, Tales of the Valdez and Copper River Gold Rush.* 1909. Reprinted 1997 (abridged), 240 pp, 5 maps, 24 photos. ISBN 1-877900-06-0.

The Valdez Museum and Historical Archives maintains a home page on the internet with additional information on the history of the Valdez area and Nancy Lethcoe's *An Annotated Database of Participants in the Prince William Sound, Valdez, and Copper River Area Gold Rush of 1898-1899.* Their web-site is located at:

http://www.alaska.net/~vldzmuse/index.html

ISBN 1-877900-09-5
Printed in the United States of American

ACKNOWLEDGMENTS

We wish to express our thanks to DeForest M. Margeson for granting us permission to reprint his father's book, *Experiences of Gold Hunters in Alaska.* It is, indeed, fortunate that DeForest and his wife Miriam chose to celebrate their 50th wedding anniversary by making a pilgrimage to Valdez to see the part of Alaska that his father so vividly described. As a result of this trip, the Margeson's donated to the Valdez Museum and Historical Archives valuable materials relating to the gold rush and brought attention to his father's book. Although it is the best published account of the prospectors who followed the Valdez Glacier route to the fabled Copper River Gold fields, it is also the rarest. Shortly after publication, the warehouse storing the books and printing plates burned destroying everything. Only a few copies survived, which helps to explain why most histories of the Alaskan/Yukon gold rushes of 1897-1900 fail to mention the Valdez and Copper River gold rush.

Fortunately, Joe Leahy, director of the Valdez Museum and Historical Archives, was able to locate and purchase a copy, which he graciously permitted us to use in reproducing the book.

As a result of publishing our *Valdez Gold Rush Trails of 1898-99,* we made the acquaintance of Larry Lashway, whose great-uncle Harry Lashway was a member of the Millard party. Because of his family's involvement in the gold rush, Larry had been looking for information in various newspapers. We owe him special thanks for sharing the articles he found on Margeson's party, the Connecticut and Alaska Mining and Trading Company.

Through Joe Leahy, we also made the acquaintance of Barry Wulff, whose great uncle, Charles Wulff, was one of Joseph Bourke's partners. Barry loaned a large set of Bourke's photographs known as the Wulff Collection to the Valdez Museum and Historical Archives and kindly gave us permission to reproduce them.

We thank these Gold Rush descendants for their helping us revive the Valdez/Copper River gold rush experience. We hope that if you have information on a relative who participated in this rush that you will contact the Valdez Museum & Historical Archives.

TABLE OF CONTENTS

PREFACE

The years 1896 and early 1897 were certainly not the best of times for the U.S. economy. Following the boom of the 80s, the nation was suffering one of those periodic economic downturns characteristic of the period's capitalism economies. U.S. monetary policy seems to have played a major role in creating this depression. Whereas, European countries no longer insisted on backing their currencies with gold, the United States clung stubbornly to the gold standard. As a result the nation's gold reserves were soon drained in paying foreign debts and the U.S. treasury found itself short of gold for coinage thus reducing the money supply. William Jennings Bryan demanded reform warning the Democratic Convention that ". . . you shall not crucify mankind upon a cross of gold." Fearing reform that might soon render their paper currency worthless, individuals began to hoard gold thus further reducing the money supply and exacerbating the crisis. During the summer of 1897, gold was very much on everybody's mind.

Although the unemployment, poverty and general hard times of the era took their toll on the American psyche, creating a certain sense of hopelessness and a lack of opportunity, many remembered the former good times and believed that the exercise of good old Yankee ingenuity, enterprise and hard work could solve the nation's economic woes. Unlike the "fin de siècle" lassitude that gripped Europe during this period, a coiled spring of optimism underlay the end-of-the -century malaise in America.

The July 1897 arrival of the *Excelsior* and *Portland* laden with gold and newly rich prospectors from the Klondike triggered the spring releasing a potential energy that would invigorate an entire nation. The image of those prospector's struggling down the gangplanks under the burden of their heavy sacks of gold had an almost archetypal appeal on the American imagination. These resourceful, self-sufficient, hard working, ordinary men and women had like our pioneer ancestor's demonstrated the pluck to challenge a wilderness and come home wealthy. They had realized the American Dream.

The newspapers now linked by telegraph and cables quickly spread this image across the nation and almost overnight, scores of thousands of ordinary people began scheming on how they might also travel to the Klondike and gather up the gold that lay waiting on almost every gravel bar. Unfortunately, few had anything but the vaguest notions of where the Klondike actually was other that it was located somewhere in the frozen north. Newspapers did little to dispel this ignorance. Rather than using the term to designate a particular river drainage system in northwest Canada, they loosely referred to the whole of Northwest Canada and Alaska as "the Klondike" and gold seekers traveling there as "Klondikers." By referring to this immense area as the Klondike "Goldfields," they seemed to imply that the entire area was covered with gold.

Those actually setting out for the "Klondike Goldfields" usually sought to inform themselves more precisely on the geography of the area. In addition to the notoriously inaccurate and exaggerated newspaper accounts, they consulted guidebooks written by purported experts. Most soon learned that by mid-1897, the entire Klondike drainage system had already been staked with mining claims; however, they were assured that the entire region which was still largely unexplored was probably rich in gold. Writer A.C. Harris advises in his 1897 guidebook that there is a feasible route to this area via the Copper River.

> J.M.C. Lewis, a civil engineer, has proposed to the Interior Department at Washington, a route from the mouth of the Copper River, by which he says the Klondike may be reached by a journey of a little over 300 miles from the coast, a great saving in distance over the other mountain routes. He says the trail could be opened at small expense.
>
> The route which he proposes will start inland from the mouth of the Copper River, near the miles Glacier, twenty-five miles east of the entrance to Prince William Sound. He says the Copper River is navigable for small steamers for many miles beyond the mouth of its principal eastern tributary. . . (Harris, p. 159).

However, anyone familiar with Lt. Abercrombie's 1884 aborted military expedition to ascend the Copper River and Lt. Allen's successful but almost disastrous ascent in 1885 would have known that the above advice was pure nonsense.

Around October of 1897, six business and professional men from Stamford and Norwalk, Connecticut, apparently influenced by Harris's advice, formed a company called the "Connecticut and Alaska Mining and Trading Association." The officers included:

D.T. Murphy, Stamford, Conn., President.
John Reynolds, Stamford, Conn., First Vice-President.
Barney Ga Taldi [sic. Gasteldi], Norwalk, Conn., Second Vice President.
Frank W. Hoyt, South Norwalk, Conn., Third Vice-President.
H.E.F. King, Stamford, Conn., Secretary.
James Hall, Stamford, Conn. Treasurer.

Secretary Harry E.F. King who wrote the company *Prospectus* enthusiastically describes the company's plans:

We are operating a co-operative mining and trading company to consist of 30 men to go to the Gold Mining Regions of Alaska, each man to contribute $400 and furnish himself with clothing suitable for the climate and pay his own fare to Seattle, which is to be the starting point: We intend to buy a schooner of about 100 tons burden, also a powerful launch, a boreing machine for prospecting, fifteen months provisions, tents, tools, medicines, blankets, sleeping bags, arms, ammunition, dynamite and drills for blasting, chemicals, forge, sawmill, large quantity of rope of all sizes, blocks, chain and anything we think will be needed for carrying on mining on a large and successful scale.

We propose to start for the mouth of the Copper River leaving Seattle about Feb. 1st. We will sail up the river as far as the rapids which are about 100 miles [sic 35 miles] up from the mouth, we shall then launch the steam launch, load it with provisions for a two weeks trip, rope, blocks, arms, ammunition, the boreing machine and ten men and start up through the rapids, the boat will not make any headway against the rapids we will send some of the men up the bank till they get above the rapids, then they can make fast one end of the rope which they will take with them to a tree or spur of rock and tie the other end to a log throw it in the river where the swift current will carry it down to the waiting men in the launch, who will speedily haul themselves up through the rapids. Once over them we have 200 miles of good navigable river, we will steam right ahead till we find the miners who have gone in there ahead of us and if they have found gold in good paying quantities we will locate near them and stake out claims and leave five or six men there and the rest will return down the river to the schooner. (Crary Scrapbook, Vol. 1, p. 112, located at the Anchorage Museum of History and Art and on microfilm).

This absurdly imaginative account of conditions the party expected to encounter once in Alaska stands in sharp contrast to Margeson's real life account of descending the Copper River near the end of his book.

King continues to describe the company's perhaps more realistic plans to "mine the miners."

The men on the schooner in the meantime will have built a good sized barge with lumber which we shall bring with us from Seattle, and have it loaded with provisions, tools, saw-mill and the whole outfit on her, if we cannot tow it up to the claims with the whole outfit on her we will make it in two or more trips. If we do not find any parties up there we will steam slowly up the river prospecting as we go with our boreing machine on every creek and gulch we see on both sides of the river. If we find gold in large quantities we intend to stake out a townsite, build a block house to live in, put up our saw-mill, establish a store or trading station, dig out some gold and send it down the river by the crew of the schooner and start them back to Seattle, show the gold and make an hurrah, boom the location, load the schooner with provisions and a number of men to work the mines.

Crowds of prospectors will rush in then we can sell them claims, building lots, provisions, lumber, tools or anything they want to buy, including medicine and then the fortune we are going to look for will [be] simply float in without much severe work on the part of the company. Now on the other hand, if we get up the Copper river and do not find gold in paying quantities (which from all reports is unlikely) we will go over the low pass at the head of the river then down the White river to the Yukon, giving up part of our provisions to the Indians to help us over the pass and sending the steam launch down the river with the crew to the schooner and let them sail back to Dyea and leave the steam launch and saw-mill there, then go back to Seattle and bring freight and passengers to Dyea until they hear from us, (which of itself will be a good paying business). . . . (Crary Scrapbook, Vol. 1, p. 112).

King continues from here to describe in equally unrealistic terms how the company would develop its backup option of continuing on to the Yukon if gold were not found in the Copper River country.

The main party will go down the White river, prospecting as we go, if we do not find gold in large enough quantities, we will go ahead into the Klondike region and get a rich claim to work on shares, when we get out some gold and the railroad is completed over the pass we will send for our saw-mill and steam launch, and a schooner load of provisions. . . (Crary Scrapbook, Vol. 1, p. 112).

King concludes his prospectus with the assurance that:

The officers of this Association are competent mechanics and receive no salary for their services, every one a hustler.

On the 15th November the Association will send a shrewd competent man to Seattle to close the deal for the schooner, launch and sawmill and attend to all other business. (Crary Scrapbook, Vol. 1, p. 112).

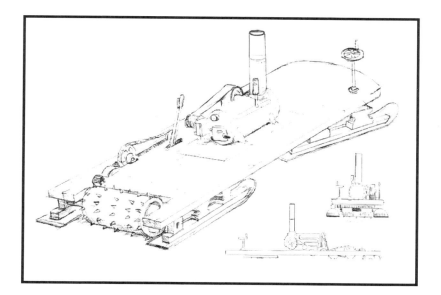

An artist's reconstruction of the steam-sled based on Margeson's description. Courtesy of the Valdez Museum and Historical Archives.

As we learn from Margeson, this "shrewd and competent man" turns out to be none other than King himself. If we can fault King and the Connecticut Company for their ignorance of Alaskan conditions, we certainly cannot fault them for their lack of enthusiasm and enterprise. However, this same peculiar blend of naive Yankee ingenuity and enterprise ended up costing the expedition dearly.

Before leaving Stamford, the company's "competent mechanics" constructed from two bobsleds and a steam engine a "steam sled" which was to carry the company's 35 tons of groceries and hardware over Valdez Glacier. Margeson describes this heavy piece of equipment being loaded on the Moonlight in Seattle and off-loaded on the ice at Valdez. He gives a graphic enough description that an artist was able to create the following sketch for the Valdez Museum.

Writing his family on March 11 immediately after arriving in Valdez, King remarks "Do not get discouraged. We are going to win out sure, and that steam sled will work out sure." (*Stamford Advocate*, March 24, 1898). However, the gap between expectations and reality could not have been greater. Addison Powell describes the steam sled's debut on the ice:

xii ■ *Gold Hunters in Alaska*

Connecticut furnished a visionary company made up of persons who were distinguished from the others by having brought a steam-sled. All they wanted was to have the right direction pointed out to them, and they would steam over the glacier, ascend the Copper River, and stampede Indians, white men and every other thing encountered. Strangers, after looking at the ponderous affair, retired to a safe distance with an expression of misgiving. When the machine was steamed up and properly directed, the owners looked at each other disappointedly, for it failed to move. They applied the full limit of steam and it stood still some more, while the joke began to settle on Connecticut. The citizens should preserve that steam sled from vandalism as an evidence of the rushers of 1898. It had the record of being the first automobile in Alaska and was never guilty of exceeding the speed limit. (Powell, *Trailing & Camping in Alaska*, 1997 edition, p. 17).

Margeson's final evaluation reinforces Powell's:

... it needed only a glance at the real situation, as it really was, to convince any practical man that it was no good for the purpose intended. This was only another illustration of the folly of constructing expensive machinery upon a theory simply, without knowing the exact conditions which will confront it in actual operation. To its originators it was an experiment, and proved an expensive one to the members of the company; but what benefit it might be to us we did not know at the time. So it was stowed away in a place reserved for it. It was expected by its builders that with it we could haul our entire outfit up the Copper River. It had cost the company, including freight to Seattle, nearly $2,000. But alas, the true conditions of things in this far-off land had never dawned on the earnest advocates of this enterprise, or the steam sled would never have been built (p. 12),

In retrospect, after experiencing all of the hardships he describes in his book, Margeson must have wondered how back in

1897 at the age of 44, he, a married man with a three year old daughter, and a successful manufacturer of steam carousels, could have been so naive as to have taken the company's fanciful *Prospectus* at face value. However, to wonder thus would be to miss the general air of unreality inspired by the Klondike discoveries. A reading of the Stamford papers immediately preceding the Connecticut Company's departure for "the Klondike" reveals the depth of this naive enthusiasm.

A Stamford newspaper reporter describing a banquet held in honor of the departing gold seekers writes:

> The Klondikers trooped into the saloon and trod over the gold-and-silver studded floor to the banquet-room in the rear with as little apparent concern as though it were but an iota to the quantities they would bring back. Some of these gold-hunters came in fur coats, sombreros, rifles, satchels and blankets, giving one the impression that a wild-west troupe had just struck town and sudden possession of the place. . . They spoke of the proposed trip and the successes which awaited them at the other end with utmost confidence. . . . (*Stamford Advocate*, 1/24/98).

An article published four days earlier describes the party's efforts to talk like prospectors: "Every member of the party has acquired an extensive vocabulary of mining terms and geographical names and uses them with great freedom in describing their future." Margeson further notes that on their arrival in Seattle some of their party attended a mining school:

> We sent three of our men to a mining school, where they would be put through a practical course of mining operations. At this school gold is intermingled with dirt, and the pupils must put the dirt through the sluice boxes, cradles, rockers, and pans, and remove the gold from the dirt. This school was held in the basement of a church, and was equipped with the same kind of machinery used in the mines (p. 12).

It is as if the party members are play acting at being prospectors and lacked a realistic appreciation of the hard work involved in reaching the Copper River and extracting mineral wealth from the Alaskan soil.

The newspapers themselves were instrumental in creating the unrealistic early expectations of the goldseekers that gold was to be found everywhere in the "Klondike." An article in the February 10th edition of the *Stamford Advocate* begins, "The Stamford Klondikers are now on the way to the Copper River and the land of golden nuggets." The same writer, describing the plight of a "gold rush widow" left behind by her gold-seeking husband, reiterates the rumors that motivated the stampeders:

> There is a pathetic side to all this Klondike business. The writer had a glimpse of it yesterday in conversation with a young wife whose husband attracted by the flowing tales of untold wealth, to be amassed so easily in that region whose rivers flow yellow with gold, permitted himself to join with the Stamford-Norwalk Klondike Co. . . . (*Stamford Advocate*, 2/10/98).

He later consoles her with "all the probabilities were in favor of 'Charlie's' swift return, in the best of health and spirits (the Alaskan climate was a wonderfully healthful one), and bringing with him as many pots of gold as though he had been at the foot of a hundred rainbows." Near the end of the description, he makes a telling comparison:

> It reminds one of those days when other "women folk" watched with hearts strained to the breaking point for the news from the front, where the blue and grey were fighting one of the bitterest and deadliest conflicts ever fought.
>
> These Klondike "waiters" should take courage, though, for (and they may thank Heaven for it) they do not know, and may they never know, the awfulness of such waiting as that. What those other women bore was truly a stupendous burden, for their news, when it came, was as apt, and often more apt, to be of death than life, while to those who are so closely connected with the gold-seekers, the probabilities are for 'a long life and a merry one.'

The writing of the period often compares the gold seekers to soldiers performing their patriotic duty—after all these men and women on the far off "Klondike" front were fighting to save the nation's economy by providing gold for the U.S. Treasury. The Stamford newspaper reports this part of the mayor's speech at the farewell banquet described above:

> By such men as you this country has been made what it is," he continued. "You go forth now carrying the Stars and Stripes into unknown lands, and may prosperity attend your efforts. Gentlemen, you have my best wishes, and God speed you. (*Stamford Advocate,* 1/24/98).

Margeson himself takes up the patriotic theme in Chapter 14:

> Much has been said since the war with Spain began about heroic sacrifice, and braving danger, all of which is doubtless true; but after the months passed on this Alaskan campaign, and seeing what I have seen, I can not think that our army and navy have given to the world any higher types of heroes—though they have been by their association rendered more conspicuous—than were developed in the wilds of Alaskan forests, or over Alaskan ice mountains (p. 88).

It is clear from the Stamford newspaper accounts that the railroads also self-interestedly promoted the Klondike fever.

> Frank A. Gross and John C. Britain, passenger agents of the Northern Pacific and Chicago, and Northwestern railroads are in Stamford to make final arrangements for the transportation to Seattle of the members of the Connecticut and Alaska Mining and Trading Association. Both of these gentlemen, aided by D.B. Carey of this city have been exceedingly active in the New England States in the interest of their respective companies, interviewing and contracting to carry prospective Klondike parties across the continent. The extent of the gold fever in the east is shown by their success. . . .

Special Pullman and diner cars were arranged for the eastern gold rushers (*Stamford Advocate,* 1/20/98) and agents from the various railroad companies escorted the group to Seattle. (*Stamford Advocate,* 1/20/98).

The ignorance of the newspapers concerning the potential for finding gold in the Copper River area was only matched by their ignorance of the area's geography. On March 24, 1898, the *Stamford Advocate* speaks of the Connecticut company's having reached "Port Valdez on the Copper River [sic]." An earlier article (1/24/'98) describes a second party of six led by Patrick Moylan which accompanied the Connecticut company to Valdez aboard the *Moonlight.* Although Margeson makes no mention of this party, the newspaper quotes letters from Moylan written while he was crossing Valdez Glacier establishing that they did indeed accompany Margeson's group to Valdez. The article states, "Three of the party who will accompany him [Moylan] are Stamford men, James Dillon, Michael Brennan and W.H. Wilson. The other two are Brooklyn men. They will accompany the Klondike party to Seattle and sail with them aboard the 'Moonlight,' to the mouth of the Yukon [sic]."

While one can be critical of the early naiveté of the enterprising Connecticut men, one must admire their adaptability in the face of emerging realities. Margeson's descriptions and other records indicate that the company did indeed carry out a number of its enterprising goals outlined in the *Prospectus.* It established a successful store in early Valdez, it set up its sawmill and sold lumber to miners, and Margeson relates how their blacksmith forge created high profits in the manufacture of ice creepers for prospectors going over the glacier. Two members of the company, Dr. Kortright and C. B. Smith both served as Trustees for the newly created townsite of Valdez. They apparently undertook these roles for civic reasons and not for the profit motives outlined in the *Prospectus.* (*Valdez Townsite Minutes:* Kortright, 4/23/98; 4/25/98; 4/26/99; 5/9-13/98; 5/16/98; 5/19/98; Smith: 10/10/98 ff).

Margeson's book, the manifest of the *Moonlight* (see Appendix), and equipment lists published in the Stamford newspaper reveal that the company directors had thoroughly researched the needs of the party and provided for a well-equipped expedition. By the time they arrived in Seattle, the party had learned of the foolishness of attempting to steam up the Copper River as outlined in the *Prospectus*. In Margeson's account, gone is talk of the steam launch (although not steam sled). Now, instead of the mouth of the Copper River, the destination is Port Valdez and the Valdez glacier route to the Copper River country.

However, once again the good Connecticut men became the victims of misinformation—this time created by the west coast newspapers, shipping companies and the U. S. military. By early 1898, it was common knowledge that the entire Klondike had been staked. West coast newspapers feeding the gold rush fever began publishing rumors of the mineral wealth of the upper Copper River noting an easy route over Valdez Glacier. For example, about the same time, the Connecticut Company was making its preparations for its voyage in Seattle, the *Seattle Times* published an article quoting an old Alaskan prospector by the name of Bayles.

> Today the Klondike country is claiming all the attention but the near future will demonstrate that the rich gold region comprises a territory fifty times as large as what is now known as the Klondike country. The Copper River country alone, with its tributaries, is over 250 miles long, by 150 miles broad, which means an area of 37,500 square miles. Gold, as well as copper, exists in abundance throughout this vast stretch of country, judging by all indications.

Some of this misinformation can be attributed to the Pacific Steam Whaling Company of San Francisco which operated the Salmon Cannery at Orca on Prince William Sound. Like the railroads, the shipping company saw a golden opportunity to enter the lucrative gold rush passenger trade and began promoting the Copper River area and the Valdez Glacier route. Most of the 3500-4000 gold seekers who arrived in Valdez in 1898 arrived aboard the company's ships. Although company officials traveling up and down the west coast touted the mineral wealth of the upper Copper River, at this time few white men had visited the area and the existence of gold in the area was pure rumor. As Margeson discovers, the area for the most part was lacking in commercially recoverable amounts of gold and was certainly a far cry from the Klondike.

The fabled ease of the Valdez route was likewise based on

other misinformation—the source this time was none other than the United States Government. In the fall of 1884, after Lt. Abercrombie had failed in his mission to ascend the Copper River from its mouth, he learned of an old Indian or Russian route that led over a glacier pass down to a large lake (Klutina) and then down a stream to the upper Copper River. He and another Lieutenant hired a Creole (half Russian) guide and several natives to show them the Port Valdez trading route to the Copper River. From here, the waters become a bit muddied as Abercrombie's description of his route fails to coincide with the Valdez Glacier trail. First, he describes an "arduous six hour climb to the base of the glacier" whereas it was only four miles over flat outwash plain to the terminus of Valdez Glacier. Secondly, he describes the glacier as trending east-west, the Valdez Glacier for the most part trends north-south. Finally, he claims that the highest point of the portage was 2500 feet and the distance to the lake 15 miles. The summit of Valdez Glacier is 4800 ft and the distance from the terminus near Valdez to Lake Klutina is 35 miles. It is little wonder that Margeson's party on hearing a rumor on the glacier that they are on the wrong route send a man down to Valdez and all the way out to Fox Island to ask Bill Beyers' native wife if they are on the correct route.

Because of the apparent duplicity in Abercrombie's report, it is difficult to explain the disparities noted above. However, we can make some sense out of Abercrombie's report if we assume that there were actually two Indian trade routes from Port Valdez to the interior—one over the Valdez Glacier and the second trending east-west and by passing Keystone Canyon by passing over Corbin Glacier down Sheep Creek thence over Marshal Pass and down the Tasnuna to the Copper River below Chitina as Abercrombie describes it and the military map accompanying his report shows it. The arduous climb to the glacier terminus fits this hypothesis. However, there is no large lake on the east side of Corbin Glacier as Abercrombie reports and the height of the pass is 3,800 ft not 2,500 ft.

Abercrombie in his published narrative describes a fog descending on the party while ascending Corbin glacier; Lt. Brumbach develops cramps and is left behind while Abercrombie and his guides proceed on ahead. From this point on, Abercrombie's account seems to be pure fabrication in order to salvage his failed mission of

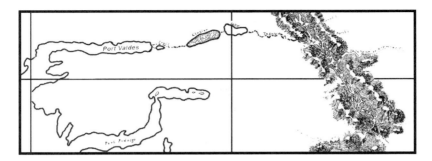

Lt. Allen's 1885 map included information from Abercrombie's alleged 1884 ascent of the Valdez Glacier Trail. Abercrombie apparently confused two different trails from Port Valdez to the interior: one, which led over Corbin Glacier to the upper Lowe River valley and via the Tasnuna drainage to the Copper River below Woods Canyon, and a second, which led over Valdez Glacier to Klutina Lake and down the Klutina River to the Copper River near the Indian village of Taral, above the infamous Woods Canyon. Since this and similar maps were used to promote the Valdez Glacier Trail, it is no wonder that Margeson's party was confused about the route (see p. 26 for a more accurate map done after Margeson's party crossed the glacier.)

establishing a military route up the Copper River. He reports the pass over Corbin Glacier as being 1,300 ft. lower than it really is and sighting the lake (Klutina?) at a distance of 15 miles from beginning his ascent of Corbin Glacier. Apparently, much to the consternation of later prospectors, Abercrombie confused the details of the two routes thinking Lake Klutina lay on the Corbin-Tasnuna route rather than the Valdez Glacier route. When he turns around before crossing the glacier, he makes up the story about seeing the lake thus confusing the later history of the route.

Considering the amount of misinformation and the gap between the Company's unrealistic expectations and the realities of the Valdez Glacier Route and the supposed mineral wealth of the Copper River Country, it is small wonder that the Connecticut Company underwent a gradual disintegration as the expedition progressed. Organizer James Hall left the company to return home immediately after the party's March arrival in Valdez. Upon reaching the foot of the glacier, Margeson is appointed assistant superintendent in charge of an advance party of 15 men. However, one of these men Lindsey Stead broke his knee while fetching water at a nearby spring. Rather than returning home, Stead rejoins the party later in the summer participating in the

Manker Creek rush. Mining records show that he is still in the Valdez area in 1901.

By the time the party, reaches the third bench of the glacier, three members withdraw to proceed on their own. Two of these turn back at the summit, while the third, Joseph Lawson, continues on to prospect the Copper River Country. Margeson meets him later during his early winter descent of the Copper River. By the time, the party reaches the fourth bench, four more turn around including Dr. Kortright who returns to take an active role in the formation of the Valdez townsite. Another of these was General Superintendent L.D. Hoy of Seattle. Margeson is then elected General Superintendent. In early June while camped on Lake Klutina, company president D. T. Murphy is forced to withdraw because of a family illness. About the same time, three more party members withdraw to prospect on their own. One of these, Captain Moyes wintered over at Grayling Creek. Mining claim records show that twelve of the remaining party members staked claims during the Manker Creek/Tonsina rush. On returning from Manker Creek in late summer, five more party members decide to return to the states. Among them was photographer, Neal Benedict, who left a photographic record and a journal of the company's expedition. Many of the photographs in this edition are by Benedict. Benedict sums up the disappointment felt by most in the following journal entry.

> We awakened to the fact . . . that somebody had been laboring under a vast misapprehension; that the idea that gold lay about on the surface of Alaska and that any innocent who had the pluck to brave the glacier and get into the country beyond would become a Midas in the twinkling of an eye, was a very much mistaken one. . . (Benedict, The Valdes and Copper River Trail, unpublished ms., Alaska State Archives, p. 10).

At this time, the party is reduced to 16 members.

By fall, Margeson and his Hornellsville comrades, Wesley Jaynes and Henry Sweet, decide to withdraw from the company as a group. Margeson resigns as general secretary. Because of H.E.F. King's harrowing autumn crossing of the glacier recorded in the book, the three decide to exit via the Copper River, having heard "that the trip had been made easily in four days." Again, the actual trip was quite different from their expectations. Although Margeson returned home after arriving at Orca, Jaynes and Sweet seem to have returned to Valdez as we have mining records for them from 1899. In all we have found records of eight members of the company who remained beyond 1898. Secretary H.E.F. King ran for the Valdez City Council in 1914.

In spite of the inflated expectations, the severe disappointment at finding no gold, and the final realization that the Copper River gold rush was nothing more than a newspaper and shipping company hoax, Charles Margeson, like others, viewed his participation in the great "Klondike" gold rush as the high point of his life. He sums this up in a final passage of the book.

> At early dawn we were on our way out of Prince William Sound, and taking the outside course, the snow-capped peaks of Alaska soon disappeared in the distance.
>
> We will not try to describe the mingled emotions with which we watched them disappear. These mountains, and rivers, and glaciers, and lakes, the hardships and perils, the pleasures and pains, the hopes, anxieties, ambitions, and disappointments which had been crowded so thickly into the past months, were all among the things that were past, but not all among the things to be left behind.
>
> Many, aye, the most of them, were to come back with us to the States as very lively memories, to go with us all the future years, to sometimes be lived over in reminiscences and story, or in dreams of the night.

After returning from Alaska, Charles and Myrtle had a second daughter, Lillian Arlene, and two sons, Morgan Eugene and Malcom Deforest. Charles and Myrtle were divorced in 1913. Charles never remarried.

Margeson went to work for Proctor Ellison Company of Elkland, PA as a shop foreman. And, just as he assumed a leadership role during his prospecting days in Alaska, so he was elected to the local school board and became the district deputy Grand Master of the Odd Fellows Fraternity. He retired in 1918, but continued to give speeches about his Alaskan experiences. In 1922 Margeson went to live with his daughter Arlene.

Deforest Margeson remarks that he remembers his father as an excellent speaker and storyteller. Like many of the prospectors, Margeson also wrote poetry about his experiences. In "Things I have Seen and Heard," he relfects on the memories that invigorated his life. The poem's last lines remind the modern reader poignantly that Margeson speaks not only of experiences that he will never relive, but opportunities that have now become rare for everyone.

Things I have Seen and Heard

There are things in life that I have seen,
That come to me with memory keen.
There are also things that I have heard,
Beyond the reach of pen or word.

I've seen men gloat when all was fair,
And breathe out curses on the air.
I've seen the same men quake with fear,
When they thought that death was hovering near.

I've listened to the ocean's awful roar,
I've watched it's breakers pound the shore
I've seen men go down to watery graves,
Dashed to their deaths by angry waves.

I've heard the grizzly's savage growl.
I've stood where wolves around me howl,
I've seen the avalanche in it's mighty sweep,
Bury comrades fathoms deep.

I've short the rapids swift and deep,
Where foaming waters plunge and leap.
I've down through rocky canyons sped,
When life seemed hanging by a thread.

I've seen the glacier's spires and domes,
    Like a thousand kings upon their thrones,
        Spread out in panoramic view,
            With miles of white and gray and blue.

I've stood on walls of adamant,
    Above the growth of tree or plant.
        And watched their peaks in morning's glow,
            Covered deep with eternal snow.

Sometimes ev'n now I backward gaze,
    To sights and sounds of bygone days.
        Sometimes in dreams I'm living o'er,
            The things I'll see and hear no more.

Charles Margeson died on January 25, 1939 at the home of his son, Morgan, at the age of 85. He is survived by his oldest son Deforest of Boynton Beach, Florida who provided us with this biographical information on his father.

According to Deforest after the publishing house produced 2,000 trial copies of *Gold Hunters*, it burned down. All the plates for the book were destroyed. He is uncertain how many copies remain but assures us that copies are rare. Prince William Sound Books is proud to be able to reprint this classic of the Valdez and Copper River gold rush.

Jim and Nancy Lethcoe
Valdez, Alaska

AUTHOR'S PREFACE

Had I anticipated writing for the public an account of my experiences when they began, I could have had something better to offer. But this thought was farthest from my mind then. Only after my return, and after most urgent solicitations on the part of many friends and acquaintances, was I induced to commence the task. Having only a limited amount of recorded items, the work must necessarily be largely from memory.

But the accuracy of memory is much accentuated by the character of the experiences through which we passed, so much out of the ordinary. Common occurrences would likely make little impression, and hardly be recalled, because of their very commonness; but experiences like these are the kind which live always in one's thought.

Hence I can assure the public that these statements can be relied upon as being accurate, and falling below, rather than exaggerating, any of the things described.

The reader will hardly need to be told that the writer has had no previous experience as a maker of books, nor has he any ambitions in that direction. But the unheard-of scramble for gold in the Alaskan and Klondike regions during the season of 1898, in which every community, almost, had its representatives, awakened such an interest in that far-away country that every newspaper item from that section was eagerly devoured by almost every person, young and old.

Of course the interest is not so intense this season, for two reasons: First, the war through which the nation is passing, and second, the large number of gold-hunters which have returned. But there is still a large number of persons in the Alaskan and Klondike regions, and the public has not lost all its interest in that section. So the writer of these pages entertains a hope that the humble offering he makes to Alaskan literature will not prove wholly uninteresting. And if this expectation shall be realized, he will be thoroughly satisfied. The fact that he sunk some eight-hundred dollars or more, and gave many months of the hardest labor of his life to this undertaking, is not altogether a regretful memory. There are many things to remember with intense pleasure, and he would hardly like to exchange the year of 1898, with all its losses and labor, for the cold dollars which they cost.

Charles Margeson
Hornellsville, NY.

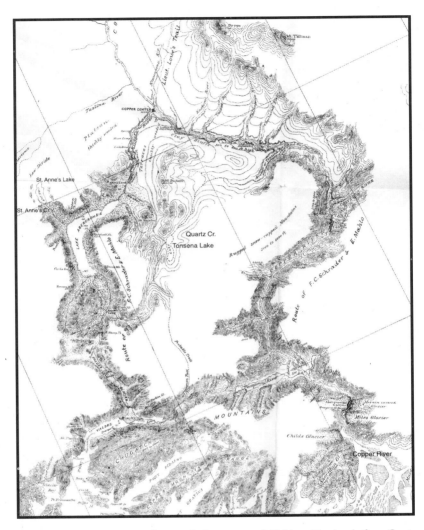

Margeson and his party began their ascent of Valdez Glacier before Capt. Abercrombie's Copper River Exploring Expedition arrived in April, 1898 to begin their survey of the route. Compare this map with the one Margeson's party probably would have used that is reprinted on page xix.

Valdez Glacier Trail from Valdes to Twelvemile Camp from the Geological Reconnaissance map of a part of The Copper River and Adjacent Territory, Alaskan Military Expedition, 1898, in Glenn and Abercrombie, Explorations in Alaska, *1899. Courtesy of the Valdez Museum and Historical Archives.*

CHAPTER I

HOW I CAME TO GO TO ALASKA

IT was during the summer of 1897, while traveling in Missouri, that I became acquainted with a man who had spent the greater part of his life in the mining districts of various countries, or rather the various mining sections of our own country, the last few years having been spent in British Columbia, where he had found good prospects, and had come to the States to form a company for the purpose of developing the discoveries he claimed to have made.

We were thrown much together, and soon became friends. For hours he sat and told me of his travels through British Columbia, and how at last he had found the spot referred to above. I was seized with a desire to accompany him; so it was arranged that we should get two other men, making a company of four, and in the spring of 1898 go together to the gold fields.

I soon left Missouri and saw no more of my friend, but kept up a correspondence with him until the early part of the winter. I then became convinced that our association together would not be harmonious, and I decided to abandon the idea of going with him.

About this time I learned that a large party of Connecticut people were forming themselves into a company to go to the Copper River, in Alaska, to prospect for gold. I immediately wrote to their secretary, Mr. Harry E. F. King, of Stamford, Conn., and by return mail I received their prospectus. He also informed me that a few places were vacant, and wound up with a request that I should join them. I discussed the question with two of my friends, Mr. Wesley Jaynes and Mr. Harry H. Sweet, both of Hornellsville, N. Y., each of whom had contracted the gold fever, as well as myself, and we decided to join this company, if upon investigation we were satisfied with their manner of doing business.

To pursue this investigation I made a trip to Stamford, where many of the members of the company lived, and attended one of their meetings, as did Mr. Sweet later on. We met many of the men, who seemed very pleasant gentlemen; so we gave them our names, paid in our money, and from this time began in earnest our preparations for our trip to the cold North. The names of the men and their addresses are as follows:

> D.T. Murphy, Stamford, Conn.
> Harry E. F. King, Stamford, Conn.
> William Williams, Stamford, Conn.
> Charles Butts, Stamford, Conn.
> James Hall, Stamford, Conn.
> William Brook, Stamford, Conn.
> Emanuel Moyes, Stamford, Conn.
> Bernard Gasteldi, Norwalk, Conn.
> Richard Voight, Norwalk, Conn.
> Joseph Lawson, Norwalk, Conn.
> Lindsey Stead, Sound Beach, Conn.
> Philip Stead, Sound Beach, Conn.
> Fred Gittoer, Cannons, Conn.
> S.J. Cone Litchfield, Conn..
> W. H. Lawrentz, Litchfield, Conn.
> John Potts, Westport, Conn..
> Henry Kitcher, Bridgeport, Conn.
> James Simpson, Bridgeport, Conn.
> Daniel 0'Connell, Glenbrook, Conn.
> T.0. Roggers. Danbury, Conn.
> Frank W. Hoyt, Norwalk, Conn.
> James Alstrum, Highwood, Conn.
> Stanley W. Gardner, Portchester, N.Y.
> Charles Priceler, Denmark, N.Y.
> Adolph Oberfeld, Booneville, Mo.
> Valentine Frickel, Orange, N. J.
> N.D. Benedict, Seville, Fla.
> J. C. Allen, Fall River, Mass.
> Charles B. Smith, Lanesboro, Pa.

Dr. Kortright, Hoboken, N. J.
L.D. Hoy, Seattle, Wash.
Wesley Jaynes, Hornellsville, N.Y.
Harry H. Sweet, Hornellsville, N.Y.
Charles Margeson, Hornellsville, N.Y.

The outfitting houses all over the country were constantly sending out printed lists of what was needed to keep one comfortable in that cold country. We selected one, and through it purchased a few articles in New York City, including sleeping bags, blankets, oil-skin suits, and underclothing; the balance we decided to wait for until we reached the coast, where it was thought we should find a better assortment to select from, and where it would be known better what would be needed.

Such frequent reference will be made in this little volume to the sleeping bag that the reader will have a more intelligent idea of what he reads about if I give a brief description of it early in the volume.

The sleeping bag is made of three separate bags, one within the other. They are about six and one-half feet long by two and one-half feet wide at the widest part,—which is the head,—and are tapered toward the foot.

The inner bag is made of heavy Mackinaw; the second, or middle, bag is of mountain goatskin, with the heavy coat of hair turned inside; and the outer bag is of rubber.

There is an opening two and one-half feet long on one side near the head, and it is through this opening that the occupant must crawl, feet first, when he wishes to turn in. When once inside, he snaps the openings together,—which openings are covered by a flap,—and thus the sleeper is entirely encased within the three bags, head and heels.

It was our custom to remove our boots, and usually our coats, and, folding them up, place both under the head of our bags, for the double purpose of raising our heads and keeping our coats and boots dry, and in some measure warm, to put on when we arose.

The weight of one of these sleeping bags is fifteen to twenty pounds, and must be carried on all tramps of any distance from camp; for one never knows when he will be overtaken by a severe storm, or how he may be delayed and be obliged to spend the night on some mountainside or in some lonely valley; and with the bag one has literally his house on his back, very much like the snail.

As the time drew near for starting, the enormity of the undertaking seemed to dawn more clearly upon us. But we had put our hands to the plow, and it would not do to look back.

A 45-70 repeating Winchester rifle was purchased for every man in the company, and cartridges enough, as we thought, to kill all the game in Alaska. But we expected to need some of them to protect ourselves against the Alaskan Indians, who, we had heard, would try to prohibit us from intruding into their domains in our hunt for gold.

The 24th day of January, 1898, was the day set upon which the company would leave Stamford, Conn. They were to go via the New York Central R. R., to Buffalo, arriving there on the morning of the 25th, where my friends and myself were to join them.

CHAPTER II

STARTING FOR ALASKA

At last the day set for starting came. The city of Stamford never saw so many people within her limits as gathered to witness the departure of this company of men,—many of them her own sons,—for their long sojourn in the land of snow and ice. Speeches were made by the mayor of the city and others, and responded to by members of the company. It was a marked day in the history of the families and friends of these men. They were going thousands of miles from home, into an unknown country,— a land full of dangers and hardships; and would they ever see their loved ones again was the one all-absorbing question which filled their thoughts.

At last the train arrived which was to take them as far as New York City. Thousands of people had gathered in the vicinity of the railroad-station, and so great was the crowd that those of the company who had lingered with their families until almost the last moment could scarcely work their way through this mass of humanity to the train. Many of the men remained upon the platform until the train began to move, when they jumped aboard, and were whirled away amid the best wishes of the great crowds who had gathered to see them off.

During the afternoon of the same day that our friends of Stamford left their homes, Mr. Jaynes, Mr. Sweet, and myself bade our friends good-by, and left for Buffalo, where we were to join our party next morning. Several of our friends accompanied us to Buffalo, and together we passed a pleasant evening, each one recounting some incident he had read in regard to this faraway country; and it was creeping along toward the small hours when we bade each other good-night, and, leaving a call for five o'clock in the morning, turned into our bunks. I seemed scarcely to have forgotten myself in slumber before there came a sharp rap, and with it came the announcement that it was five o'clock. Oh, dear! Five o'clock so soon! And the train which was to bring the larger part of our company was to arrive at six. So we had no time to waste. We had barely time for a hasty breakfast when the train pulled into the station, and with many good wishes, and some hurried hand-shakes with our home friends, we stepped on board the car which was to be our home for the next five days.

The first few hours were spent in getting acquainted with the members of the company whom we had not met before. At Suspension Bridge the train made its first stop, and here the Connecticut delega-tion ate its first breakfast on board the train; and as they brought out their large and well-filled lunch baskets, it became evident that we were not the only ones who had supplied themselves with an abun-dance of good things to eat during our journey to Seattle.

About noon, at a little town called Sand Run, our train ran into a fearful blizzard, and it became so severe that we were obliged to stop until another engine with a large snowplow was brought to our assistance. It was with difficulty that the two engines could pull the train through on schedule time. While it was gloomy enough outside, it was far from it inside. We had two violins, an accordion, a piccolo, and several harps, and many of our men were excellent singers; so with "music, song, and story," the time passed very pleasantly.

As the reader is more or less familiar with the country through which we passed, I will not attempt to give a detailed narrative of our trip to the coast but will only give a few incidents which served to break the monotony, and give an added spice to what would otherwise become tedious.

We encountered huge snowdrifts all through Nebraska and North Dakota, but with the aid of the snowplow we pushed through admirably. By this time our train was made up of twelve coaches, and was loaded exclusively with "Klondikers" who had been picked up along the way. As the train sped on, the enthusiasm which was mani-fested at the start was somewhat abated. The people had become so accustomed to seeing train-loads of gold-seekers that they took it as a matter of course, without comment.

The larger part of our company knew each other so well that jokes were always in order, and it was a rare thing for many hours to slip past without some one breaking out on somebody else in such a way as to create great merriment. Here is a sample: One night it seemed to have been thought best to place a man on watch, to look after things in general, and to guard against "hold-ups," while the others slept. So one of our number was selected for this duty, armed with a well-filled revolver, and stationed at the front of the car, where he would be ready to do battle with any one who should attempt to board the train with unlawful purposes.

About two o'clock in the morning the sentry became drowsy, and was soon fast asleep. In this condition he was noticed by one of his friends, who was always around when there was a chance for a bit of fun. So quietly slipping up to the sleeping sentry, he quickly removed all the cartridges from his revolver without disturbing him, and left it where he had found it. Then going through the car, he told several of the boys whom he thought would enjoy the joke, and together they suddenly yelled, "Murder! Robbers!" Instantly the sentry was on his feet, with revolver grasped tightly, ready for any deadly work which might be on hand. The first intimation that he had been sold came to him in the roar of laughter which greeted him, as he flourished his weapons and struck his warlike attitude. Still further humiliation greeted him when he found that he was about to make war with an empty gun. It is needless to say that he had to "set up" quite freely when we reached Seattle.

The scenery through the celebrated "bad lands" of Dakota was new to most of our party, and was greatly enjoyed by all.

While passing through Wyoming, we discovered a large pack of coyotes out upon the prairie, half a mile from the train. Many revolvers were emptied at them from the car windows while the train was in motion, but of course the bullets fell far short, and the simple coyotes were scarcely aware that they were the objects of interest or attention.

It was in Wyoming that the boys got their first sight of Indians, on their native soil. They attracted much attention, and many wondered if they looked much like the Copper River Indians, about whom we had heard so much, and from whom we expected trouble before reaching the interior of Alaska.

While passing over the Rockies, the scenery was varied and beautiful, and was much enjoyed by us all. The snow was five feet deep, but the road had been kept open by its frequent trains, and ours was not seriously delayed. Notwithstanding the storm through which we had passed, we were but five hours behind schedule time. While upon the summit, we could look far down the mountainside, and see the road over which we were to pass, but to reach which we had to wind around for many and many a mile. Coming down into the valley, the snow had entirely disappeared, and grass was as green as in New York State in May. Truly a sudden transition from winter into the lap of summer; but which is the mountain traveler's experience.

The foothills were covered with a dense growth of low scrub pine, in which, we were told by men at the little stations through which we passed, there was an abundance of grizzly bears and other large game. The mountain streams were clear as crystal, and were said to be filled with brook trout. This made our hunters and fishermen anxious to get off and spend a few days trying their luck in these wilds of the West; but as we were not on a hunting or fishing expedition, they kept right on with the party. Nothing worthy of note occurred, and we rolled into the station at Seattle on Saturday afternoon at two o'clock. The platform was crowded with people, and as we looked from our car window, we were reminded of circus day in a small town.

Before our train came to a full standstill, our secretary, Mr. Harry E. F. King, came into the car. We were of course glad to see him, and he seemed glad to be among his friends again. Some three weeks before, our company had sent him on to, Seattle, as our advance agent, to purchase a schooner, large enough to transport ourselves and our goods to Alaska. He had purchased for us the schooner *Moonlight*, for which he had paid $2,500. He had written us about the purchase before we started, and he had made arrangements with draymen to take our belongings from the station to the schooner, so we had not to bother with that. He conducted us at once to the vessel, but one block from the station.

The morning papers had given out that a carload of Yankees, all belonging to one company, were on their way from the Eastern States, and would be in Seattle that morning to "outfit" for the Klondike; and I believe that every outfitting store in the city had its representative at the station to meet us. Every step of the way to the boat we were confronted by hotel porters, hack-drivers, baggage and express men, steamboat runners, and by men representing all manner of outfitting stations. Besides these, there were scores of men who had the latest improved mining machinery, without which it would be worse than useless to go to the Klondike. Each man tried to outdo his neighbor in setting forth the advantage of outfitting from his particular house. We took all the cards they gave us, and I doubt if a bushel basket would have held all the printed matter our company had thrust upon them during this little trip of one block, from the station to the boat. Talk about enterprise! If you don't find it in Seattle, especially when they have a "fresh arrival" of good proportions to work upon, then I wonder where, in all the round world, it can be found? Don't think, however, that we were annoyed by the evidence we saw of its existence. On the contrary, we enjoyed it. Why, weren't we Yankees ourselves? and hadn't the most of us come from old Connecticut? And if the enterprise of the East had taken on some new elements, and some Western enthusiasm, why, it had more room, and it could.

Of course we all took a lively interest in the schooner which was to carry us and our belongings so many miles over the great Pacific. We stepped down upon her deck and examined her closely. Her masts were large and strong, and her rigging and sails were new. Going below deck we saw that her timbers were strong and well put together. She was built in 1890 by a Norwegian fisherman. Her length was 69 feet; beam, 28 feet; depth of hold below deck, 7 feet. She had a gross-tonnage of 71-81, with a net registered tonnage of 68-22. She had too much beam to her length to be a fast sailer, but we cared little for that if she would only prove to be a safe one. We spent the afternoon looking over the schooner, and noting the changes that must be made in fitting her for sea. Enough of this sort of work was found to keep us busy for several days.

The next day was Sunday, and while walking around town we

were surprised to find that a very large number of business houses had their doors wide open, and were doing business the same as upon other days. It was said that there were forty thousand "transients" in the city, and judging from the crowds everywhere in the streets, the number was not overestimated. Ship-loads were leaving every day for the gold fields, and many times a day there were new arrivals on every incoming train to take their places.

On Monday morning committees were selected to take charge of every part of the work of fitting our vessel, and they were instructed to push it to completion as quickly as possible. The next few days were busy ones, and in three days she was ready for her cargo.

By this time the various committees on supplies had completed their work, and the goods were rapidly hauled to the wharf and stowed away beneath the deck of the *Moonlight*. During the week we put on board over forty tons of provisions, hardware, and other goods; and last, but not least, the celebrated steam sled, which had attracted so much attention in old Connecticut and other States. While bringing it from the station to the boat large crowds gathered around it, and many were the comments passed upon it. Having said this much, it may be well to give the reader a brief description of this wonderful piece of mechanism; for it is quite safe to say that nothing was ever before made like it, or will ever be again.

The steam sled was primarily two very heavy bobsleds, the runners of which were five by seven inches, and six feet in length, and made from timber which had a natural crook for the bend. The shoes were of half-inch steel, eighteen inches wide. They were both seven feet wide on the ground, and were fastened one behind the other, with a space between of four feet. Upon these bobs was fastened a platform made of two-inch plank. It was eight by sixteen feet in size, and so arranged that the foremost sled was left free to turn right or left as desired. In the center of this platform sat a ten-horsepower boiler (upright) and an eight-horse reversible engine. In the rear of the sleds was a large cylinder two feet or more in diameter and four feet long, filled with heavy spikes. A large amount of gearing connected this cylinder with the engine,

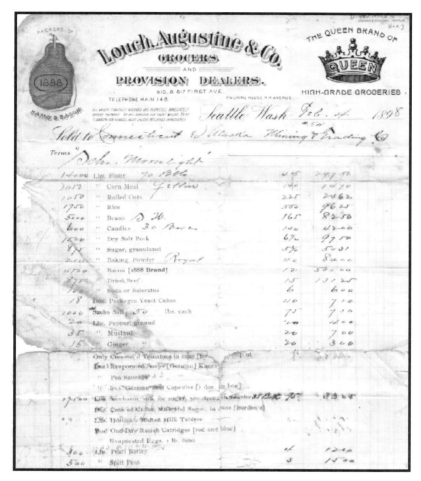

Manifest of the Moonlight, *Margeson Collection, Courtesy of the Valdez Museum and Historical Archives.*

and when in motion the spikes upon this cylinder were supposed to dig into the snow or ice over which it was to be driven, and it was supposed that this machine was of sufficient capacity to draw large loads of our goods over the glaciers, instead of packing them or drawing them by hand.

But it needed only a glance at the real situation, as it really was, to convince any practical man that it was no good for the purpose intended. This was only another illustration of the folly of constructing expensive machinery upon a theory simply, without knowing the exact conditions which will confront it in actual operation. To its originators it was an experiment, and proved an expensive one to the members of the company; but what benefit it might be to us we did not know at the time. So it was stowed away in a place reserved for it. It was expected by its builders that with it we could haul our entire outfit up the Copper River. It had cost the company, including freight to Seattle, nearly $2,000. But alas, the true conditions of things in this far-off land had never dawned on the earnest advocates of this enterprise, or the steam sled would never have been built.

In addition to all the other goods taken on board the *Moonlight*, our company took along a large amount of goods as "trading stock" to barter with the Indians. A Seattle paper, in speaking of this, said: "The innate thrift of these Connecticut men is seen by the great amount of stuff carried aboard their vessel, and if the Indian maidens in the vicinity of Copper River are henceforth found decorated with rings of purest brass and beads of gaudy color, it will be known that the *Moonlight* shed its rays of Connecticut hope upon that far-off country, and took in exchange some gold."

We sent three of our men to a mining school, where they would be put through a practical course of mining operations. At this school gold is intermingled with dirt, and the pupils must put the dirt through the sluice boxes, cradles, rockers, and pans, and remove the gold from the dirt. This school was held in the basement of a church, and was equipped with the same kind of machinery used in the mines.

Our company was incorporated under the laws of the State of Washington, with a capital stock of $75,000. It may seem strange to some that we should have been incorporated in Washington rather than in Connecticut, where the larger part of its members lived. But the fact is that the laws of Connecticut regarding incorporations are such that they could not be lived up to in our case. They require that the greater portion of the stock of a company must be held by residents of the State, and every member of our company was expected to be absent from the State as much as

"During the week we put on board over forty tons of provisions, hard—ware, and other goods; and last, but not least, the cele—brated steam sled, which had attracted so much attention in old Connecticut and other States."

Photo by Neal Benedict, Messer Collection, courtesy of the Cook Inlet Historical Society.

two years. It was thought advisable, therefore, to organize under the laws of Washington, which are different in this respect.

On Monday, February 7, we engaged the services of Captain Leonard Bare to take command of the vessel, and also engaged four sailors and a cook. Events proved that our selection of captain and crew was wise, but I can not say as much for the cook; for as soon as we reached Port Valdez, it was unanimously agreed that he should never again ship as cook on the *Moonlight*. Next day we loaded eight thousand feet of lumber and took on board supplies for our trip.

Everything was now ready, and we were only waiting for our clearance papers to get under sail. These were somewhat delayed, and it was not until the afternoon of February 9, that word was sent out that papers were received, and that the *Moonlight* would set sail in twenty minutes. In an instant all was confusion on board, and as some of the boys were on the shore, we began to feel anxious lest they might get left. Couriers were sent out to hurry them up, and in about fifteen minutes our secretary mounted the boom, and in a loud, clear voice, called the roll, and every man responded to his name.

CHAPTER III

SETTING SAIL ON THE GREAT PACIFIC

THE little tug *Mayflower*, which had been engaged for the purpose, steamed alongside, and began making preparations for towing us out into the sound. Handshaking with the new-made friends of Seattle was soon over, the lines were cast off, and the little tug was slowly towing us toward the great ocean.

As the boat left the wharf, three rousing cheers were given for the State of Connecticut. Then one of the boys climbed into the rigging with his piccolo, and began playing *The Girl I Left Behind Me.* And amid the waving of hats and handkerchiefs, the little tug bore out to sea, not only the high hopes of the brave men on board, but the hopes of a larger number of wives, mothers, sweethearts, and kindred of every tie, besides the warm friendships of a lifetime in the far East. For it must be remembered that a sail on the ocean was a new experience to almost, if not quite, all our company. And it will not be thought strange if there were in that crowd of apparently cheerful, hopeful men an undertow of very serious feelings, and an inward questioning as to what might be the outcome of this long trip; and whether it would prove to be simply a phantom chase, with only blasted hopes for a recompense, or whether all this cheerful and hopeful crowd would again set foot on the shore we had just left. Well if this was so, we think there are few who have had any similar experience, who have it in their hearts to chide our boys with "weakness." The sound was soon reached, sails were hoisted, and as they caught the light breeze, the little tug gave the signal that she had turned us loose. As we moved slowly down the sound, darkness settled over the waters, and the lights from the city made a sight not soon to be forgotten. Then there came a calm, and for two hours we lay almost absolutely motionless.

Our musical members brought out their instruments, and made melody on the waters, and the time passed pleasantly and quickly away. About nine o'clock all hands of us turned in for the night and slept away our first night at sea. During the night the wind sprang up, and daylight found us sixty miles from Seattle. There was a small amount of work to be done below deck, and after breakfast a few of us, myself among the number, went down to do it. But we were soon glad to get on deck again, for the sea began rolling, and shortly after, as we looked from man to man about the deck, we noticed that a pallid look had settled down upon many faces. Soon fully one half of the boys were lined up along the side of the vessel, and seemed to have a frequent and uncontrollable desire to lean over the rail, and make suggestive motions to the sea.

Before eleven o'clock the sea calmed down, and by noon all but a few were able to do ample justice to dinner. During the afternoon we were favored with a light breeze, which moved us along at a four-knot rate. We passed large quantities of floating sea-weed, and fished out some beautiful specimens of it; and often the remark was heard, "Oh, if I could only have that at home." But alas! such a thing could not be, and the find would be thrown back into the sea. Just at dark the little breeze which had been moving us along leisurely during the afternoon, died out entirely, and left us in a dead calm. We had been twenty-four hours out, and the captain informed us that we were seventy-five miles from Seattle.

When daylight came next morning, we found we were ten miles nearer Seattle than we were the night before. The tide had borne us backward while we lay helpless without wind for our sails. The men began to get uneasy because we were making so little headway, and it seemed to be the forerunner of general fault-finding.

About eleven o'clock a whale was sighted a short distance from the vessel, and immediately the men forgot their uneasiness, and all stood watching the great black fish, as he moved, lazily along on the surface of the water. We also saw large numbers of seals. The little animals would raise their sleek-looking heads about a foot above the surface, look at us a moment, and then disappear.

Dinner-time found us still becalmed, and drifting with the

tide. About two o'clock a breeze was felt from off our lee bow. It struck us lightly at first, but began to increase until our boat was scudding through the water at a terrific rate, and in another hour the sea was lashing itself into foam, and the spray was dashing over our deck. Near Cape Flattery is a small harbor, in which it was our captain's intention to take refuge, if possible, until the worst of the gale had spent itself; but about this time the wind changed to dead ahead, and all hope of reaching the harbor was cut off, and all we could do was to tack back and forth across the straits. As darkness came, a gloom seemed to settle down on all the men, and no one had much to say. One by, one they went below and turned in, but very few of them could sleep for the fearful pitching of the vessel.

About midnight the wind ceased, and the boat was left with flapping sails, and rolled helplessly in the trough of the sea. Toward morning of February 12, a slight breeze sprang up, and we made a little progress until about nine o'clock, when we were again becalmed. The sea continued to run high, though we had no wind, and the boys were mostly lined up against the ship's rail, and looked as if they had lost every friend they ever had. During the forenoon we drifted several miles backward toward Seattle, but before noon a light breeze filled our sails, and we moved slowly along past the island of Vancouver, whose shores were in plain view, and seemed heavily timbered with spruce. At two o'clock the wind freshened from the north, and we were able to take a westerly course. The wind kept on increasing, and when darkness set in, we were fast leaving Cape Flattery and Vancouver.

As the wind increased, so did the waves, and our little craft was soon being tossed about on the water like a veritable cork. By eight o'clock the wind had become a gale, and the waves were rolling mountain high. All hands were ordered below except the sailors. We obeyed, and went to bed; but it was almost impossible to remain there. Not a man could sleep for the tremendous rolling of the ship, and the roaring of the seas as they swept the deck from bow to stern.

At 3:30 Sunday morning, February 13, the wind lulled somewhat, but the sea continued to run high. After daylight the wind changed to dead ahead, and we were obliged to change our course to the southeast. The captain said that he wanted to get at least one hundred miles off Vancouver Island before he dared to lay his course, for fear another storm might come which would drive us on the coast. Our boat was supplied with neither keel nor center board, which kept us from running close into the wind, and we were often drifted far out of our course.

All day Sunday, with scarcely wind enough to fill our sails, we kept on a southeast course. The sea continued to run high and all but five or six of the men were sick, many of them so bad as to be scarcely able to raise their heads from their bunks. The cook was too sick to attend to his duties, and two of our number, Messrs. James Simpson and Daniel 0'Connell, took their places in the cook's galley, and during the balance of the trip they were conspicuous characters around that part of the ship. Often the appetites of the men who had eaten nothing for some days would be tempted by some delicacies prepared by these boys and were thus enabled to eat something, when otherwise they could not have tasted food at all. They were not sick during the entire voyage and were thus able to render valuable service to very many who were too sick themselves to do anything, even to the caring for their own needs.

Sunday night came and went with no breeze, and we drifted around with our sails flapping. Monday morning, February 14, brought rain and fog, but about nine o'clock the sails began to tighten in a light breeze from the northeast, and we were able to take a westerly course.

Many of the boys began to be homesick. Could they have been now set back at home, nothing in the wide world could have tempted them to again embark on such an enterprise. But it was too late for such thoughts now; we were all in it, and must get out as best we could, if indeed we were ever to get out at all.

After dinner the wind freshened, and the sea again began to run high, and by four o'clock it had become a gale. All sails were taken in, and we were running before the wind with bare poles. The captain ordered just enough sail hoisted to give the boat steerage-way, and then lashed the wheel to position.

The roaring of the sea was something terrible. Our little craft was first on the crest of a great wave, and then far down in the trough of the deep. Monstrous waves would sweep clean over her, and all night long it seemed impossible that any boat could live the storm out in such a sea. But our little schooner proved that she was stanch and true, and in a master's hands, for daylight of our sixth day out dawned upon us, and she was all right. Not a timber was strained, not a leak had occurred, not a spar had been shivered, but at least twenty-five of our boys were too sick to leave their bunks, and very few could say that they were well. Some were anxious to turn back and give up all they had put into the enterprise, but the majority were for pushing ahead. The wind was too high all day to carry much sail, and all our ship could do was to hold her own.

Toward night the weather turned piercing cold, and a blinding snow-storm set in. This greatly increased our discomfort, for our stove in the forecastle smoked so that we could have no fire. We had twenty-six men forward and sixteen in the cabin, and our quarters were very much cramped because of having so much freight, and so many sick. I assure the reader that our position was not an enviable one. None of us thought we were having a "picnic."

Our captain here took his reckoning, and found that we were one hundred and seventy miles from Cape Flattery, and one hundred miles from Vancouver; so we had at least two weeks more of sailing to expect before reaching our destination. Night set in with the wind still in our teeth, and blowing too hard to carry much sail. All Tuesday night we rolled and tossed about at the mercy of the waves. I shall never forget the appearance of the sea, as I stood on deck just before dark, clinging to the rigging, and looked out over the angry waves. The water looked to be as black as ink. Over the bow of the vessel I saw great waves that appeared to be forty feet high come rolling toward our little craft, and it seemed as if the next moment we must be swallowed up in the great deep. But the little thing would be lifted up, for a moment poise on the crest, and then make another plunge far down, till it seemed as if we were surely to rise no more. The awful grandeur of that scene can never be effaced from the memory of those who saw it. Men may strut and swell on land, and even on the water may feel somewhat important while the sea is smooth, and all things go well; but at such a time as this, when nature's elements are marshaled as for deadly combat, and set themselves to a test of strength, one can but feel that he is less than a speck of the great creation. If the reader ever feels that he or she is quite large, just take such a trip as this, and you will be cured for all time—that is, if you are sensible.

Wednesday, February 16, brought no change in the wind, and we seemed to be just drifting about, and getting no nearer Valdez, which was our destination; and this caused some of our men to lose courage, and long for an opportunity to get once more on solid land. I felt sorry for some of the boys, for they looked the picture of despair. Others seemed full of hope and courage, and apparently never had a thought of turning back.

About noon the wind, which had been blowing a gale for two days and nights, changed slightly to the southwest, and a severe hail-storm set in. We put up our sails double-reefed, and turned the ship northwest, and were able to make some progress against the terrible seas which sometimes swept our decks. Toward night the wind gathered strength for another reign of fury, and fury it was. There was little sleep on board the *Moonlight* that night, though the men were tired enough to sleep almost anyhow; but the vessel rolled so that one could scarcely keep in his bunk.

Thursday morning, February 17, our eighth day out, came, and brought with it rain and some snow. The wind was still very high, and our little ship labored heavily, as she took plunge after plunge into the seething waters.

During these rough days it was amusing to see the boys come on deck, and make for the cook's galley for breakfast or dinner, as the case might be. There it was dished up, and every man, on receiving his plate of food and cup of coffee, would make an attempt to reach the hatchway; but before he could reach it, the ship would make a sudden lunge, and he would go sliding across the deck, and in his eagerness to catch on to something would lose his coffee or his plate, and perhaps both. He was considered a lucky fellow indeed if he got below with his meal, without having it drenched with salt water.

All day the wind blew from the southwest, and we made but

"Few were able to eat any breakfast; but about ten o'clock the sun, which we had not seen in several days, made its appearance, and the sea so far subsided as not to break over our decks."

Photo by Neal Benedict, from the Messer Collection, courtesy Cook Inlet Historical Society.

little advance because we were obliged to carry so little sail, and the great waves kept beating us back. The sick seemed generally better, although a few were still very ill. In the afternoon again the wind and waves increased so much that by six o'clock it was next to impossible to get on deck for any supper, and few of the men made any effort in that direction. All night long the wind and sea were something terrible. About midnight our jib-sail was carried away, and the terrific plunging of the vessel caused great fear among all on board; and not without reason, for it did seem that no boat, great or small, could outride such a sea. Our little boat would come up out of each immersion, seeming to shake herself as the waters poured off her deck, thus getting ready for another plunge. All on board passed the night without sleep, but toward morning the wind subsided a little, enabling us to put out a little more sail, although the sea had the appearance of a great body of boiling suds. Few were able to eat any breakfast; but about ten o'clock the sun, which we had not seen in several days, made its appearance, and the sea so far subsided as not to break over our decks. Wasn't it a treat? After such a fearful baptism as ours had been for so many days, old Sol never shone with such an interesting face, so we thought.

The doctor took this occasion to order every man on deck for pure air, and the well ones had to assist the sick. A few were very sick, and they were carefully wrapped in blankets and carried up, and seemed much benefited by it.

A large gooney, or albatross, had been following us all day, picking up what crumbs or other things had been thrown overboard, and we made an effort to capture it with a hook baited with pork and attached to a long line. Several times he made the effort to secure the bait, but the boat was moving too fast, and we had to see him fall out of the chase.

At two o'clock the wind died out entirely, and by eight o'clock we were again in a dead calm. The sun had shone brightly all the afternoon, and a great portion of the time had been spent by the men on deck, and had been greatly enjoyed, as the sea had become comparatively smooth. Soon after eight o'clock, however, the wind began to freshen, and we were not destined to enjoy a long rest from the power of the blast. By midnight we were in it again, but our ship was running before it, and was behaving splendidly, and we had the satisfaction of knowing that for every hour passed at that rate we were eleven miles nearer Copper River.

On February 19, at the first dawn of day, I went on deck, and the sight that met my gaze was one never to be forgotten. The wind was blowing at a tremendous rate, and the great seas, fully forty feet high, were chasing each other over the deep. Each wave was crested with foam as white as snow, while underneath was a line of deep sea blue. It was terrible, yet grand to look upon.

Our sick were much better, and the thought that we were being rapidly borne toward our destination caused the men to be more cheerful and hopeful. Nevertheless it was the most terrible day we had yet experienced. One thing was in our favor, we had plenty of sea room to run before the gale. Great seas washed our deck all day, our jib-sail was again carried away, and one of our boats torn loose from its fastenings. Several seas swept over the cook's galley, which was six feet above deck. Two men were cutting steak from a quarter of beef, when a great wave broke over the ship, carrying men and beef across to the opposite side, and one of them was considerably injured against the rail. After this, no one but the sailors was allowed on deck while the storm lasted, and they worked much of the time in water. We were now all cooped up below deck, and with so much sickness, the air soon became very foul. Our discomfort was increased by being obliged to hold fast to something all the time to keep from being flung to the opposite side of the vessel.

Night came on, but brought us no relief. As darkness settled over us, the storm seemed to increase in fury, and all night long the demons of the deep and air appeared as if turned loose. To us, shut up below, and listening to the creaking of the ship, the roaring of the wind, and the rushing of the great waters, and unable to see anything of what was going on, it seemed as if every lurch we made was a downward plunge from which we might not arise. Small wonder that the night was one of great tension on nerves so wholly unused to such a life, or that the hours were weary ones while we waited for the morning.

At last Sunday morning dawned. It was our second Sunday at sea. About ten o'clock the storm began to abate, and by two in the afternoon the sails were flapping idly, with not a breeze to fill them. The ocean continued to run high until evening, when it calmed down considerably, and we passed a comfortable night.

Not having a clergyman on board, the day was not observed very religiously; for not more than one or two of our company, and none of the crew, pretended to be Christians. Yet after so many days of glaring into the very jaws of death, as we had done, we can but think that a feeling akin to gratitude must have taken possession of every thoughtfulf person. Of course the boys slept soundly after having been kept awake so long.

Monday morning, February 21, dawned bright and clear, and all hands were allowed to come on deck to enjoy the occasion. We were soon scudding along under full sail and made fair progress. The whole of that day and the following night we had favoring winds, and we were nearing our destination at a commendable rate.

Tuesday, February 22, was bright and much colder. We had hoped to be at our journey's end by this time, but instead we were but a little over half way there. By favorable weather, however, we hoped to cast anchor in Prince William Sound within another week. All day Tuesday we made but five knots an hour, and that not in a direct course, for we had head winds, and were compelled to bear to the southwest. That night gave us a placid sea, and the boys got a fine night's sleep.

Wednesday, clear and cold. After the sea became smooth, the sick improved rapidly, and soon, with but few exceptions, were able to eat heartily.

The captain took his observations, and reported that we were 595 miles from Port Valdez. This was a disappointment to us all, for we supposed we were much nearer.

In the afternoon considerable excitement was created by the sighting of four whales, and all hands were on deck in short order. They remained several minutes on the surface, then suddenly disappeared. All that afternoon we moved but little, there being scarcely any breeze. Shortly after nine that evening our sails were moderately filled, and by eight o'clock next morning we had made sixty miles.

About noon, Thursday, the wind died out, and we were left in another calm, which lasted nearly all night. Toward Friday morning the wind freshened, and we made five knots an hour, when we were again left in a calm which continued until after midnight.

When we awoke Friday morning, February 26, we were running under full sail, and making six miles an hour. The breeze continued all day Saturday, and during the night had increased so that by Sunday morning the sea was again running very high; but we were making good time, and that was becoming a chief consideration.

Sunday morning, February 27, our eighteenth day out, the weather was much colder, and all hands passed the greater part of the day below decks. We were now nearly opposite Prince William Sound, though still far out at sea. The wind was not favorable for running in, and so we had to tack back and forth until the wind should favor us. We had now been so long at sea that the men hailed with delight any breeze which helped us along toward our journey's end. But a contrary wind gave us no pleasure, you may be sure; and when we were obliged to turn the bow of our boat away from Prince William Sound for no one knew how long, we were not in the best of humor.

Monday, February 28, was a beautiful day. The sun shone brightly, and the thermometer registered at 40° above zero. On this day we caught our first sight of the snow-capped peaks of Alaska.

Toward night a dense fog arose, and not daring to remain so near shore, we put out to sea, and ran before the wind several hours; then coming around, we began tacking toward shore.

On Tuesday morning, March 1, the fog cleared and we again sighted land, and ran up to within forty miles. We could see that the whole country was covered with snow, and looked desolate enough; but we felt to welcome almost anything for the privilege of setting foot once more on solid ground. In the afternoon we ran up to within about twenty miles of shore and could see the lay of the country quite distinctly. Ahead of us was an immense glacier, the first we had seen, and we spent much time examining it through our glasses. There were places upon the foothills where the snow must have been over a hundred feet deep, for we could see that the tops of the tall spruce trees were only just visible above the snow line.

Five or ten miles out from shore stood several rocks towering hundreds of feet above the waters, like sentinels, to guide the mariners into Prince William Sound, which was thirty-five miles distant. Storms are of such frequent occurrence along this coast that we dared not pass the night near shore, so we turned about and put out to sea; and it was well we did, for about eight o'clock the wind increased so that we had to reef our sails. The sea became very rough, and we passed another sleepless night. We kept out to sea until four o'clock Wednesday morning, when we put about and stood toward shore. When daylight appeared we were out of sight of land; but soon the snow-capped mountains were sighted, and we ran toward them all day, keeping as close into the wind as possible. When within twenty miles of shore we found that we had gained but twenty-five miles during our sail of one hundred and fifty miles, and were yet ten miles below the entrance of the sound. We were therefore obliged to put out to sea for another night.

On the morning of March 3, after sailing all night, we found ourselves in sight of shore, but becalmed. The entrance to the sound was directly in front of us. About noon the wind began to blow off shore, and we began to tack back and forth, hoping to be able to enter the sound before dark. The clouds looked threatening, and it was evident that another storm was gathering, so we decided, if possible, to run into McLeod's Bay, which is a small body of water near the entrance to Prince William Sound. The entrance to it was narrow and rocky, and it was with feelings of relief that we got inside, and dropped anchor in a splendid spot near the shore of Montague Island.

The little bay was three miles long and one mile wide, and surrounded on three sides by mountains one mile high. It was a most dismal-looking place. The shores were heavily timbered with spruce, which extended up the mountainside for half a mile, at which point the tall trees were nearly buried in snow, and above that point nothing was visible but snow and ice.

Having been so long at sea, our drinking water, which was carried in barrels upon deck, had become stale and really unfit for use; so we decided to go on shore, and, if possible, replenish our stock. Night had set in, and it was very dark, yet we lowered a boat, and five men went on shore. They had been cooped up so long on the vessel that as soon as they set foot on shore, though it was dark, and the snow deep, they at once began to run races up and down, and acted like schoolboys just let out of school.

As they were walking along the beach in their search for water, they were greatly surprised to come across a small cabin, nestled close to the foot of the mountains under some large spruce trees. They had no light to examine it, so they returned to the boat and secured a lantern, and returned to the cabin. They found it deserted, but it gave evidence of having been occupied within a few weeks past. It was built of poles, and covered with thin slabs split out of spruce logs. On the ground in one corner was a bed made of spruce boughs, large enough to accommodate two or three men. A few slab shelves were up around the wall. Scattered around upon the ground floor, and also about on the exterior, were the bones of some animals which had evidently been used for food by its occupants. Who had built it,—whether white men or Indians, or for what purpose, in such a desolate and deserted spot, was of course only a matter for speculation. It proved this, however, that we were not the first human beings who had put foot on this inhospitable shore.

After looking over their find until satisfied, they continued their search for water, and were rewarded by finding a clear stream coming down from the mountainside. They filled their buckets, and hastened back to the vessel, and had only been there a few moments when the storm broke on us in terrible fury. It is needless to say that we were exceedingly glad for such a shelter as our little bay afforded; for had our schooner been out in the sound, she must almost certainly have been driven upon the rocky coast, and what our fate would have been we shuddered to conjecture. As it was, she dragged anchor, and before morning grounded on a sandbar; but by taking advantage of the tide, we succeeded in getting her in deep water again. But the weather looked so threatening that we decided to remain the day out in our snug little harbor. We made an effort to catch some codfish, but found the water too shallow for them. We did, however, catch some very fine starfish the finest that I had ever seen. One of these measured nearly eighteen inches across, and had seventeen points. Some of these would have been valuable additions to any collection, and I confess that I coveted at least one to put with my other specimens at home; but alas, we had no place to accommodate such curios, so we threw them back into the bay.

During the day we all went ashore and treated ourselves to a good wash in fresh water, the first in twenty-two days; for on board the schooner we were only allowed to use fresh water for culinary purposes, so had to wash in sea-water during the entire trip.

Toward night the wind changed, and we weighed anchor, and before dark were sailing up Prince William Sound within seventy-five miles of Port Valdez. Saturday morning, March 5, found us becalmed, thirty-five miles from our destination. We sighted a steamer, evidently on her way to Seattle, and hoped she would pass near enough to us to give us an opportunity of sending mail back to the anxious ones at home; but we were disappointed, for she passed us at a distance of three miles, and was soon out of sight.

All that day and night we lay becalmed. To be so near our destination, and not to be able to make any headway, was indeed trying; and the men becoming uneasy, we decided to attempt to tow the schooner into Port Valdez by hand. So we lowered our two boats, and manned each with four men; then fastening long lines from the boats to the schooner, we began to row. This was indeed slow work, but we were able to make one mile an hour. The time given each crew to row was an hour, and then a fresh crew would take their places. This was kept up until afternoon, when we sighted a steamer headed for Port Valdez. A hurried consultation was held by the company's directors, and we decided to engage them, if possible, to tow us into port. So we signaled them, and they came alongside. We offered them $50 to tow us in, but they asked us $150. We were not inclined to pay that price, so we let them go on their way.

Just after dark the wind came up from ahead, and the channel through which we had to pass was narrow and rocky. We disliked to run back over the few miles we had worked so hard all day to gain, so we hoisted sail, and determined, if possible, to work our way up against the wind; but had just got under way when a blinding snowstorm was upon us. All night long we tacked back and forth across the narrow channel, and I considered it the most dangerous night we had passed in all our rough experience of so many days, from the fact that the channel was full of sunken rocks which were not marked upon the chart. The night was dark and stormy, and neither our captain nor any of our crew had ever been in this port before.

Next morning we ran into a little bay, and dropped anchor. The wind died away, but the snow continued to fall until noon, fully a foot having fallen since the night before. There seemed to be several arms to the bay, and we had no idea which would lead us to Valdez; so, as soon as anchor was dropped, a boat was manned and sent out to investigate. At noon they returned with the discouraging report that they had gone several miles up one arm of the bay—the one we considered most likely to be the right one—and found ice, but no signs of port.

Manning the boat with a fresh crew, they set out for another tour of investigation. During their absence it was suggested that we try our hands again at another attempt for cod. The poor success which had crowned our previous efforts was still fresh in the minds of the men, and but few thought it worth while to try; but one of the company, who was a persistent fisherman, thought differently, and baiting his hook with fat meat, threw it over the side of the boat and almost before the others could think, had a twenty-one-pound cod flopping on the deck. This, of course, set the boys wild, and soon a half dozen lines were out; and in a single hour a half ton of wriggling codfish was landed on the little schooner. Then some of the boys, wishing to eat a meal cooked on Alaskan soil, took some provisions and a fresh cod and went ashore, which was but a few hundred yards away. There they made a camp fire and cooked and ate a supper which they declared was fit for a king.

Just before dark we heard several shots in the direction of the camp fire, and looking through a field glass we saw one of the boys emerging from some bushes bearing a huge hedgehog. It was the first game killed in Alaska. They had seen his track near their camp, and following it a little way, found the fellow perched on a limb a few feet above their heads. Two of the boys emptied their revolvers at him, but he sat there winking and blinking at them, as if he considered his position the safest one in the field. The boys became desperate at their bad marksmanship, and placing their weapons back in their holsters, went down to the beach, and gathered their hands full of stones, and with these returned to the tree. Their aim proved better with rocks than with bullets, and soon Mr. Hedgehog was dropped from the tree and quickly dispatched with a club. The victorious hunters returned to the schooner bearing their game, and seemed to feel as proud as if they had bagged a grizzly bear.

It was late at night when the boat returned, and reported that they had found Valdez, and that it was seventeen miles away. They said that many people were there getting their goods off the ice, where they had been unloaded. We remained where we were until morning, when there was a slight breeze; and hoisting all sail, we moved slowly toward our destination. About noon we were again becalmed, and again began towing the schooner along by hand as before. All the afternoon we pulled hard, and when night came, were four miles from the edge of the ice where we were to unload. We ran in behind a small island, and anchored for the night. The weather looked threatening, and after dark it began to snow. In the morning there were ten inches of snow on deck, which had fallen during the night. After clearing the deck of its snow covering, there being no wind, we resumed our work of

"About one o'clock we drew up along the edge of the ice near where the steamers were unloading. Going back some distance from the edge, we cut a hole in the ice, and hooked our anchor into it, and our boat was thus held firmly in place. . .So we began in earnest the work of un— loading the schooner."

Photo by Neal Benedict, from the Messer Collection, courtesy of the Cook Inlet Historical Society.

towing. We could see men unloading a steamer on to the ice, and hundreds more were hauling their goods back a mile to Valdez. At a distance of four miles they looked much like a procession of ants, all busy at work. We could plainly hear the howling and barking of dogs as they were being put to work hauling the goods back off the ice to a place of safety.

It was thought advisable by our company that a general superintendent should be appointed to have charge of and give direction to, all the work when we should land; so a meeting was called, and Mr. L. D. Hoy, of Seattle, Wash., was elected to that position.

About one o'clock we drew up along the edge of the ice near where the steamers were unloading. Going back some distance from the edge, we cut a hole in the ice, and hooked our anchor into it, and our boat was thus held firmly in place. The ice was about eighteen inches thick, and covered with three feet of snow, while away from the ice the snow measured eleven feet on the level. When we saw this condition of things, many of us said in our own hearts, "The steam sled is a failure;" yet we were hoping that it might be of some service, after we reached the solid ice of the glaciers. We must wait and see. So we began in earnest the work of unloading the schooner.

First our sleds were taken off, fourteen short and ten long

ones. The short ones were four feet long, and intended for one man each; and the long ones were six feet long, and intended for two men to use. The lumber was the first thing unloaded from the deck, and piled up, a little being hauled off the ice. Next came the famous steam sled. This was so heavy that it was no easy task; but "many hands make light work" proved true in this case, and soon the ponderous thing was resting some distance from the schooner, on snow and ice.

About a mile from where the schooner was anchored was a piece of timber containing two or three hundred acres, and running down through this was a clear stream of pure water. In the edge of this timber, and near this little stream, were about one hundred tents, clustered together, and others were being put up. This unique camp—for it was about that—presented a scene of unusual activity. Some were tramping down the snow, preparing a place to set up their tents; some were cutting tent poles, and others cutting firewood, while others were getting their dog teams ready for hauling their goods up to the foot of the glacier, which was five miles away.

A few of our company went up and selected a site for our seven tents, tramped the snow, and set a stake in the center, on which was written, "This space taken." This would prevent any other party from entering upon our labor, if they were to come on the ground befor° we returned. Next morning we discovered that the great weight of the goods unloaded from the steamers and from our schooner, with our lumber and the steam sled, had caused the ice to settle so that eight inches of water had overflowed the surface, which caused us great fears for the safety of our goods, which had been unloaded. So we decided not to place any more on the ice until we had removed what was already there to a place of safety. The day was spent getting our lumber on shore, and getting our steam sled farther from the edge of the ice. On the following morning we began the unloading of our thirty-five tons of groceries and hardware. The water being so deep over the ice, we had to carry our freight some distance back before it could be loaded on the sleds. This made our work not only slow, but much more laborious. It took us three days, toiling late and early, to get our goods to a place of safety in case the ice should break up, as there were signs of its soon doing.

During the second day we saw that we must get our steam sled off the ice, or risk its going to the bottom; so attaching strong lines to it, the whole company drew it to a place of safety on shore.

Toward noon, the third day, the ice had become so rotten that large cakes became detached, and would settle under us into the water as we passed over them with our goods. The last half day of this work was attended with great danger, for had any of us broken through with our loads, drowning would have been the almost certain result. I remember well one time, while drawing my load, that I stepped on a large cake, which broke into several smaller ones with my weight, none of which were large enough to support me, and leaving my load, I sprang lightly from cake to cake until I had reached firm ice; then, getting the assistance of some of my companions, we carried lumber and bridged the spot sufficiently to bring over the balance of our goods. We were all glad when the last of it was got off, for there was scarcely a man but could relate some thrilling adventure during his work over the rotten ice. No accident happened, however, for which we were really grateful.

After the work of the day was over, many of the men sat up far into the night, writing to the loved ones at home; for the steamer was to start on her return to Seattle next day.

Several of the passengers who came on this boat, when they saw how deep the snow was, and what difficulties lay in the way of getting our goods over the glaciers, became discouraged, and returned with the same boat. Their golden dreams of finding a fortune in Alaska had been rudely dashed to atoms in those few days of rugged experience in barely getting on shore, and they were returning, discouraged and disheartened.

Since we anchored our boat to the ice, the weather had been unusually fine. The sun had shone brightly every day; but as we took our last load from the boat, clouds began to gather, and the barometer gave promise that a storm was brewing The captain informed us that the safety of the schooner might depend on getting her away from her present situation before the storm broke, for the ice would be broken up, and she might be drifted into and crushed by the grinding mass. So a few hurried preparations were made, and we bade adieu to the schooner *Moonlight*. Just before dark she weighed anchor, and as there was a good breeze blowing down the bay, all sails were hoisted, and in a few hours she must have been far out on Prince William Sound. Upon her return trip she encountered several severe gales, one of which carried away part of her sails, and otherwise damaged her, but she made the trip back to Seattle in the unprecedented time of nine days.

During the night upon which she sailed away, a storm came on, and in the morning the great field of ice over which we had carried and drawn our goods, had broken loose from shore, and was ten miles down the bay.

About this time the steamer *Valencia* came into the bay, with six hundred passengers for Valdez. The captain had left the lighter or small boat, at Orca, thinking they could run up to the ice and unload as we had done; but when they discovered that the ice had gone out, and that there was no way to unload their large amount of freight, they decided to land passengers and goods at Swansport, which was four miles down the bay. This the passengers would not agree to, although the captain gave orders to commence the unloading of freight; but when the ship's crew went to lift the hatchways, they found a body of armed men guarding them. This caused a parley, which lasted several days; but the men were firm, and the outcome of it was that the captain sent for his flatboat, and passengers and freight were landed at Valdez, as they desired. This made about fourteen hundred men that had landed at this port within the past few weeks, and other boats were daily expected with hundreds more.

Our tents were up, and we had begun camp life in earnest. We were better provided for than many of our fellows, for we had lumber to floor one half of each tent, so that we were not obliged to spread our sleeping bags on the cold snow at night, as most of them must. The glacier seemed but a short distance away, but we were told that it was five miles. This seemed to us impossible, but when we loaded up our sleds, and joined the procession which was on its way thither, we soon found that the distance had not been exaggerated.

Two trips a day to the foot of the glacier and return was a hard day's work, but it was done by many of the men; others, who were not disposed, or were not able, making but one trip daily.

Situated about two miles from our camp, in another piece of timber, and about the same distance from the glacier, were a few rude log cabins and several tents, the former having been built several years. One man had lived here nine years. This was Valdez proper, and the old Indian trading post. Years ago many Indians came over the glacier during the latter part of winter, bringing sled-loads of valuable furs and articles of their own manufacture, and traded them for beads, brass trinkets, gaudy-colored clothing, provision, and such other articles as seemed to strike their fancy. It was on one of these trading excursions that one of their women had her feet so badly frozen that, when they returned, she was unable to walk. So she was abandoned, and left to freeze or starve, as the case might be; but her situation was discovered by a white man, who took pity on her, and took care of her. She soon recovered, and they have since been living together. This man moved soon after to an island in Prince William Sound, where he became engaged in raising blue foxes; and by the disposition of his valuable furs has been making a good living. The island is now known as Fox Island. A few weeks after our arrival at Port Valdez, this same Indian woman gave us some valuable information, of which I will speak later on.

A few years ago a trading post was opened up at Cook's Inlet, and to this place the Indians could come to do their trading without crossing the glacier. As a result, but few Indians now come to Valdez, and the trading post here, which a few years ago was in such a flourishing condition, is now a thing of the past. When the great Alaskan gold excitement sent thousands to land at Valdez, the old site of the town did not seem to them to be the proper place for a town, so a new site was selected, which was near deep water, and seemed to be a more suitable place for a thriving village.

Upon this new plot streets were laid out, and during the summer many log and some frame buildings were erected, so that in the fall of 1898 Port Valdez was a village of three hundred people. The first boat-load of gold seekers that landed at Valdez took up their quarters at the old town site, among whom was a party of four men, who camped, or tented, together, and were doubtless in some sort of partnership. They failed to agree, for some reason, and almost every day found them in difficulty. One day, in a fit of anger, one of the party, Doc. Tanner by name, rushed to the tent in which his three companions were, and drawing his revolvers shot two of them dead, the other escaping at the rear of the tent. Of course this was more than the law-abiding people could stand, and the murderer was soon placed under arrest; a judge and jury were selected, and the culprit was brought forth to trial.

The case was such a plain one that the trial was short, and he was soon found guilty of murder in the first degree, and sentenced to be hanged immediately. They did not stop to erect a scaffold, for a limb of a tree would serve the purpose; so he was taken out to a large tree, and a rope placed around his neck. All this time he seemed the least excited of any man in the company, and talked about the horrible crime as if it were a matter of no importance, and jokingly disputed with the men as to which limb of the tree he was to hang from. The limb being selected, the rope was thrown over it, and he was quickly drawn up, and thus paid the penalty of his bloody deed. Now the old town of Valdez bears the name of "Hang-town."

This event caused the people to believe that crimes of greater or less magnitude would be committed from time to time, with such a conglomeration of citizenship during the year, that it should be impressed upon the minds of all who landed at this port that no crime, however small, should go unpunished. So a meeting of the miners was called, and a Law and Order Committee was selected, whose duty it should be to make the laws, and affix the penalties thereto. As there were large quantities of goods of all descriptions piled up along the trail, it was only natural to expect that stealing would be most prominent among the cases to be dealt with in the history of crime along the trail; so particular stress was placed upon this crime, and severe penalties attached. It was to the effect that if a man stole ten dollars or less, his crime was petty larceny, and the penalty for which was to have his goods confiscated, and he was to be driven from the country; but if his theft was ten dollars or more, it was grand larceny, the penalty for which was hanging.

". . . I never knew a place in any civilization where the principles of honesty were so thoroughly carried out as upon the Valdez and Copper River trail. Goods could be left upon any part of the trail for an indefinite time without their being molested."
Crary Collection, B62.1.2944, The Anchorage Museum of History and Art.

This punishment may seem to many as too severe, but when it is remembered that one's provision was his life in a country like this, it was regarded that a person who stole that was as truly taking life as if he went to the tent and killed the owner outright.

Whether the making of this law was really necessary or not, I can not say, but I do know of no single instance in which it had to be enforced; and I never knew a place in any civilization where the principles of honesty were so thoroughly carried out as upon the Valdez and Copper River trail. Goods could be left upon any part of the trail for an indefinite time without their being molested.

If a man found anything on the trail which had been lost by another, he would place it upon a stick, stuck up in the snow, beside the trail; or if a man found among his goods a sack or box which belonged to another, he would put up notices in two or three conspicuous places, to the effect that a box or sack bearing such a mark had been found among his stuff, and invite the owner to call and get his property.

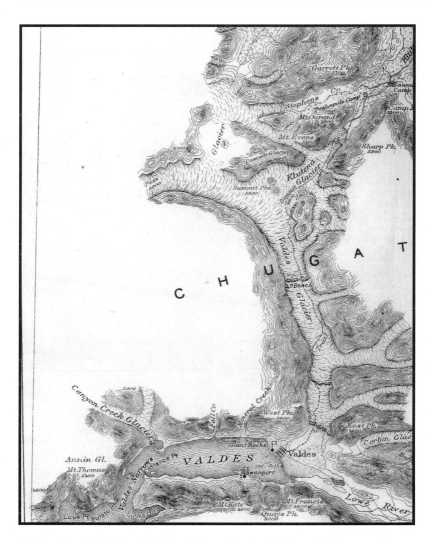

Valdez Glacier Trail from Valdes [old spelling] to Twelvemile Camp on Upper Klutena [old spelling] River. *From the Geological Reconnaissance map of a part of The Copper River and Adjacent Territory, Alaskan Military Expedition, 1898, in Glenn and Abercrombie,* Explorations in Alaska, 1899. *Courtesy of the Valdez Museum and Historical Archives.*

CHAPTER IV

STARTING OVER THE GLACIER

WE had been landed at Valdez about a week. Our goods had been brought from the shore, and placed in a cache near our camp, and from there we hurried them along as fast as possible to the foot of the glacier. We had brought along a good supply of goods to be disposed of at as good advantage as possible after our arrival at Valdez. These goods must remain more or less exposed to the elements until we could get a place prepared for them, so a large body of men were set to work to build a store. To prepare the foundation for this building was no easy task, as there were ten feet of snow to be removed to get at the ground, and much of this had to be drawn away on our hand-sleds to get it out of the way. The sills and sleepers had to be hewn out from standing timber, and carried by hand to the building spot, where they were placed upon temporary abutments of wood, which was to do till summer made it possible to replace them with something more enduring. About fifteen men were kept at this building, and in a few days our store was completed, and the goods which we had intended for immediate disposal placed within.

"Our goods had been brought from the shore, and placed in a cache near our camp, and from there we hurried them along as fast as possible to the foot of the glacier."

Camp Valdez. Photo by Neal Benedict, Messer collection, courtesy of the Cook Inlet Historical Society.

"We had taken along a blacksmith's forge and tools, and a large quantity of sheet steel, and having two black–smiths in our company, we set one to making 'creepers.'"
Note the blacksmith holding a creeper.

Photo by Neal Benedict from the Messer Collection, courtesy Cook Inlet Histor–ical Society.

About this time one of our men, Mr. James Hall, of Stamford, Conn., became convinced that he could not stand the work and hardships which must necessarily be encountered during our stay in Alaska, and he decided to return to the States. It was not long before he had an opportunity, and bidding the company good-by, he set out for home and friends.

Many of the men who preceded us had begun the ascent of the glacier, and there came a great demand for "ice-creepers." We had taken along a blacksmith's forge and tools, and a large quantity of sheet steel, and having two blacksmiths in our company, we set one to making "creepers." These found a ready sale at $3.50 per pair, and he could easily turn out twelve pairs a day. Thus we were able to reap a large revenue off the labors of one man. There was a large demand for provisions, also, and all kinds of goods used in camp life. Bacon brought 25 cts. per pound; pork, 20 cts.; flour, $10 per hundred; beans, the same; and every other kind of provisions brought from two to four times its cost. The seventy-five mattresses with which we had furnished our schooner before leaving Seattle, at a cost to us of 75 cts. each were taken off at Valdez, and quickly sold at $3.50 each. There was also a great demand for tents and tarpaulin, but unfortunately we had but a small amount of these for sale. A tent costing $7 would sell quickly for $25. The lumber used in building the store could have been sold for a profit to us of $70 per thousand. Goods sold so rapidly that it was thought best to send a man immediately to Seattle to purchase a new supply for our store, so Mr. John Potts was sent back to purchase a stock of goods, and hurry them on with all possible dispatch.

On the 18th of March we moved three of our tents to the foot of the glacier. At this time there were from seventy-five to one hundred tents located there, and every day some men who had gotten their goods a mile or so up the ascent, were moving out and going on up farther for a new camping place; but scarcely would they vacate their old camping ground before some one who was not fortunate enough to have gotten his goods as far along as his neighbor, would drop on the place he had vacated, and set up his tent. This would save the new comer the labor of leveling off and packing the snow on his so-called village lot. Besides, he would often be able to get a location in the heart of the city, when otherwise he might have to go far down the trail; for in this village of tents there was but one street. In order to hold a lot it was only necessary to set a small stake, on which was written the word "Taken," and no one would attempt to occupy it as long as the stake remained.

When our three tents moved to the foot of the glacier, our company was thus divided; and our superintendent being kept busy at Valdez, I was appointed an assistant, having charge of the three camps of fifteen men. The wind blew so hard nearly every day that we were greatly hindered in our work. The snow was constantly drifting into the trail so that we were compelled to load light, and many days could make but one trip a day, when with a good trail we could as easily have made two. All the wood which we must provide for our long trip over this mountain of ice grew from four to six miles from its base. There was a vast quantity of dry timber in these woods, but it was fast being taken, and we thought it wise to devote a few days to supplying ourselves with wood for our cold journey; so, going to the woods, we spent several days felling trees, cutting them into six and eight feet lengths, and hauled them on our hand-sleds to the base of the glacier, where they were piled up for future transportation up the mountain of ice.

One morning as we were on our way up the trail with our loads of wood, we saw several men examining something, and on coming up to them, saw that it was the track of some large animal. It had come to a cache of goods and carried away several sides of bacon. This was more than the owner could stand, so getting a few men to accompany him, they put on snowshoes, armed themselves with Winchesters, and started in pursuit of the thief. They followed his trail several miles, and came up to him in a small piece of timber, where a well-directed shot laid him dead. It proved to be an enormous black lynx, the fur of which was very valuable; so the man considered himself well paid for his stolen bacon. Not every thief pays for his thieving with his own hide; if he did, goods would not often change hands that way.

About one fourth of a mile from our camp was a spring of clear, sparkling water which came bubbling out from under an immense rock. From this spring we carried all the water used while camping at the foot of the glacier. The trail leading to this was in many places steep. One day as one of our company, Mr. Lindsey Stead, of Sound Beach, Conn., was bringing a pail of water, he slipped and fell, breaking his kneecap. Our physician, Dr. Kortright, was at Valdez, five miles away, so we summoned another physician, who reduced the fracture, and the patient was made as comfortable as possible. He was then placed on a hand sled, and taken to Valdez, where comfortable quarters were arranged for him at the store, and he could receive constant attention from our own doctor. This was a serious blow to him, for he had been among the most ambitious to get into the interior of Alaska. It was thought that in two or three months he might be able to join us, but at the expiration of eight months he was still at the store, and very lame, though able to be about.

The first funeral after we reached Alaska occurred shortly after we reached the foot of the glacier. It was that of a man who had drawn his goods about two miles up the mountain of ice. He had undertaken to draw too heavy a load, and had ruptured a blood vessel, dying in a very short time. He belonged to the Masonic fraternity, and the Masons took charge of the funeral. He was wrapped in his blankets, and placed on a hand sled, which was drawn by six Masons, four others walking behind, holding to a line attached to the sled to hold it back as they descended the glacier. Next came about sixty Masons in line, and then several hundred miners, who, to show their respect for their dead companion, had quit work to attend the funeral. His last resting place was near the foot of the glacier by a little cluster of bushes growing close to the base of the mountain which towers thousands of feet above it.

Shortly after this another accident occurred which cast a gloom over the entire camp. There was a party from Wisconsin camping near us. Among them was a bright young man who, a few weeks before, had left his home, gaining from his parents only a very reluctant consent to his coming. This company used two tents, one for cooking and eating purposes, and the other for a sleeping apartment. There came a day when the wind blew too hard to work on the trail, and these men concluded to take this time to clean and oil their guns, which were packed in a box in their sleeping tent. They all sat in their cooking tent, as was their custom, when this young man requested a boy about sixteen years old, belonging to the company, to go and bring his gun. No one seemed to know that the gun was loaded, but soon a report was heard, and the young man leaned over against his companion, as they sat together, and said, "I am shot." The bullet had passed through the tents, and through his body. The wound was at once pronounced fatal, and though everything possible was done for him, he lingered two days and died. Next day a grave was dug in the frozen ground beside that of the man whose funeral has already been described, and a large company of miners followed him to his lonely resting place.

Before giving an account of our ascent of the glacier, it may be well to attempt a brief description of this great mountain of ice, about which so much has been written, and upon which we toiled so hard for sixty days, getting our goods over. The glacier is a large body of ice, lying between two mountains, which rise above it on either side from three to five thousand feet. Its length is thirty miles, and it has an average width of three miles. It is twenty miles to the summit, and it has an elevation in this distance of five thousand feet. And down the opposite side to the valley beyond it has an average descent of three hundred and twenty-five feet to the mile. Upon the summit of this ice mountain, from early fall until

First Bench, Valdez Glacier. "Our camp was at the foot of the first bench, which is about sixty yards in length, and so steep that it was necessary to cut steps in the ice to get up." Photograph reproduced from first edition.

late in the spring, fierce snowstorms rage almost every day, so that during the winter the snow accumulates to a great depth. During the summer the snow melts from the ice, and leaves the whole surface of the glacier one mass of yawning crevices, many of them hundreds of feet deep, which make its passage impossible in the latter part of summer and early winter, and extremely difficult and dangerous at any time but the latter part of winter, when they become so filled with snow that there is little danger in passing over them. In making the ascent there is a series of five benches to go over, the last of which is called the summit.

Our camp was at the foot of the first bench, which is about sixty yards in length, and so steep that it was necessary to cut steps in the ice to get up. At the top we drilled a hole in the ice, into which was set a post, attaching to it a pulley, through which we passed the end of a rope. Then attaching each end of the rope to a sled, about ten men would climb to the top, get hold of the rope and empty sled there, and come down the incline, drawing the loaded sled up, carrying from six to eight hundred pounds of goods at a load. Other companies were doing this also, and often several lines were being worked on the same bench at once. We would frequently have to wait several minutes for others to get out of our way, and they as often waited for us; but I do not recall a single instance in which any hard feelings or angry words were indulged by any of the various companies who worked upon these benches. I consider this a remarkable thing, when men were working so hard as almost naturally to become irritable when very tired; and if anybody doubts that they were very tired, he would soon be convinced by a day or two of actual trial of it. Many hundreds were passing and repassing upon this narrow trail, and necessarily were often much in one another's way. And each one seemed ever ready to lend his neighbor any assistance needed, when it was possible for him to do so. Quite often some man's sled would slide out of the trail, and tip over with its load in the snow, but the first men passing him would get out of their harnesses, help him back into the track, and assist him to reload. This same spirit seemed to be manifested during our entire trip over the glacier.

About the time we were getting our goods up the first bench, a large steamer came into the bay, and landed at Valdez over six hundred Swedes. These men began to work night and day, and they soon had their complete outfit at the foot of the first bench. There were so many of them that they were in one another's way, and during the next week the ropes on the first ascent, or first bench, were kept busy day and night. My tent was exactly in front of the trail leading up to the steep wall of ice, and not more than fifty feet from its base. During their first day's work at this place there was considerable excitement caused by a loaded sled breaking away when near the top, and coming down at lightning speed, ran over two or three men, quite seriously injured one of them, and crushing through one corner of my tent, where only a few moments before some of our men had lain asleep. We ran out and cautioned the men against such carelessness, and going in, sat down to breakfast; but before the meal was finished we were again startled by the cry, "Look out for a runaway sled!" and running outside, were just in time to see it upset and spill its contents in the snow just above us. This brought a sharp reproof from our boys, with another promise on their part to be more careful in the future.

These people were ignorant in the use of a rope, few of them understanding how to tie a safe knot, and we never rested easy while camping below them. Several times in the night we were awakened by shouts, which we knew meant a runaway sled, and we would raise up in our sleeping-bags and almost hold our breath as we would hear the downward rush of the sled, expecting every moment that it would come crashing through our tent, until we could hear the splurge of the scattering goods as they went hither and thither over the snow; then we knew our peril was over for that time, and would cuddle down to resume our sleep. But we could never sleep quite easy as long as these people worked nights.

From the top of the first bench to the foot of the second was a half-mile, and the ascent here was so great that one hundred pounds made a full load. The second and third benches were so close together that they might almost be called one, and they together measured one thousand five hundred feet at an incline of forty-five degrees. It was over this that the trail ran on which we were to haul our goods, and here again the rope and pulley were put to service. We stretched about seven hundred and fifty feet, and after getting our stuff all up the second bench, the rope was carried ahead, and the next pull brought them upon the top of the third bench. Following the goods came our wood, which was drawn up in the same manner.

About this time four of our company's men, Joseph Lawson, J. B. Allen, Adolph Oberfeld, and Valentine Frickel, became desirous of severing their connection with the company, to pursue their search for gold by themselves; so a satisfactory adjustment was affected with them, and they withdrew. But they continued the ascent of the glacier with us until the summit was reached, when Frickel and Oberfeld abandoned the scheme, and returned to the States. Allen and Lawson pushed on into the interior, and passed the summer upon the Klutina and Copper Rivers.

While working upon the third bench of the glacier, the view down the valley was unobstructed. Valdez was in plain sight, and the trail leading to the glacier was marked by a long line of dark moving objects which were mere specks in the distance. They were men, pulling their goods up the trail on sleds as we had done. Ev-

"While working upon the third bench of the glacier, the view down the valley was unobstructed." Photo by P.S. Hunt, Crary Collection (B69.13.23), The Anchorage Museum of History and Art.

ery steamer or schooner which came into the bay could be plainly seen, and almost every day brought a new addition to the already large crowd of gold-seekers. We couldn't help thinking that if all of us were to strike luck, then indeed must Alaska be full of the shining metal.

We let our camps remain at the base of the first bench as long as possible, so that we might save the wood which had been already drawn up for future use farther on; but the time came when it was too much of a task to walk so far to our work and back, so we moved them a mile above the top of the third bench. From the top of this bench to the foot of the fourth was twelve miles. All this distance the ascent was gradual, the elevation being not more than one hundred feet to the mile. We could take one hundred and fifty pounds to a sled, and if the weather was good, could make a round trip a day; but if the weather was threatening, we would unload at what was known as fivemile camp, which was less than half way.

Many men had dogs, which were of great assistance to them. A large dog would haul as much as a man, and often we would see a load of five hundred pounds drawn by three or four dogs. Frequently we would see a man and his dog hitched in the harness together, the dog walking along beside his master, and, where the road was good, pulling the entire load.

Many of these dogs were but half fed and when they became weakened by overwork, some of their masters would beat them in a most brutal manner, and it was no uncommon thing to see a dog drop dead in harness beside the trail. A few men were so cruel to their dogs, and beat them so much, that a committee was sent to wait upon them; and they were ordered to stop beating them, or they would be dealt with by the indignant miners. This seemed to have the desired effect, for from this time on the poor dogs fared better.

It was no uncommon sight to see a horse plodding along up this trail of twelve miles, hauling ten or twelve hundred pounds to a load. We were seized with a desire to possess one ourselves, but they were scarce and in great demand, and the price asked for the few which were for sale at all was enormous. One day a man came down the glacier who was the proud possessor of three. He had got his goods on the summit, and would therefore sell one. The animal was gaunt and thin in flesh, and looked as though his friends had long forsaken him. But the price asked for him was $300. We held a consultation and decided to buy him. We had a new demand on us at once—something to feed our animal; so a delegation was hurried off to Valdez for hay. A few bales were secured, costing at the rate of $100 per ton. The trail from Valdez to the foot of the glacier had become so soft that the horse could not be used upon it, so the hay was drawn by hand to a place where it could be reached by the horse. The harness used on our horse was not such as an Eastern harness-shop would like to hang outside to advertise its work, for it consisted of a collar made out of old sacks, a pair of hames made from two crooked sticks fastened together at top and bottom with pieces of rope, and a pair of rope tugs. Could it have been preserved, and returned to the States, it might have occupied a prominent place in some museum; but it is doubtful if either horse or harness will ever leave Alaska.

Our next need was a stable for our horse. To make this we dug a hole in the snow eight feet square and seven feet deep, and covered it with a tarpaulin. The drifting snow would soon form a nice warm roof, but without the tarpaulin cover, would soon have filled up our stable, and buried our horse. An inclined path down into it was shoveled out, and in this way our animal was protected from the storms which so often swept the glacier.

Conspicuous upon the trail were several women who had gone to Alaska with their husbands, and nearly every day we would meet them trudging along through the snow, pushing with a stick placed against the rear end of a loaded sled, to assist their husbands with their loads. A few of them did no work but to cook for their husbands and care for the tent in their absence, while others went out and toiled on the trail all day, doing the work of a man.

One day, just on a tour of investigation, one of these women walked from Valdez to the foot of the glacier, then up the mountain of ice fifteen miles, and then returned to Valdez, making a round trip of forty miles in one day, over a trail which, at its best, was bad enough. This same woman walked over the glacier several times during the summer, and in her search for gold penetrated almost as far into the interior as did any of the men. I remember one day during the summer, while on a prospecting trip, meeting one of these women dressed in male attire plodding along beside her husband on a steep mountain trail thirty-five miles from camp.

Another instance which illustrates the pluck and endurance of these women, was one who, even before we began the ascent of the glacier, made herself conspicuous by her hard labor on the trail, from early morning until late at night; and many days when it was so stormy that few men cared to expose themselves to the elements, this little woman and her husband could be seen trudging along with a heavily loaded sled. All the way over the glacier her physical endurance was the wonder of all who saw her. For months they had toiled and got their goods over to the head waters of the Klutina River, toiling almost night and day. The snow left them, so that sledding was impossible; and as they could not well whipsaw lumber for a boat, they purchased one, paying for it $60. Loading their goods into this boat they started down the swift-flowing stream, and had gone but a little distance when their boat was capsized, and all their goods which they had labored so long and hard to draw over the great mountain of ice were lost. It seemed as if this was enough to break the spirit of most people, under such circumstances but having a little money left, they watched for opportunities where provision could be obtained cheap, and were soon in possession of another complete outfit.

There was one woman on the trail who felt at home at almost

any kind of work, but seemed at her best when participating in some exciting adventure. She could guide a boat down the swiftest mountain stream equal to an Indian, and she seemed anxious for an opportunity to shoot the great Klutina River rapids. She could handle and shoot a rifle with the dexterity of an old hunter. She always joined her husband in his hunting excursions, and many birds and animals were brought down by her unerring aim. She was good-natured and always jolly, having a pleasant word for every one she met, and seemed greatly to enjoy the kind of life she was living.

Our goods from Seattle arrived while we were camped near the top of the third bench, and we were sent for to assist in unloading and getting them up to the store. Leaving our work at the glacier, we spent two days at Valdez, in which time our goods were safely housed. Returning again, we resumed the arduous work of hauling loads over this long stretch of uphill trail to the foot of the fourth bench, and it used up many days before they were all there.

About this time Mr. L. D. Hoy tendered his resignation as general superintendent of the company, which was accepted, and I was elected in his stead. We had now the larger portion of our goods at the foot of the fourth bench, so I moved my tent and one other there, while the three remaining tents went on to the foot of the summit, five miles beyond the fourth bench.

At this time Messrs. S. J. Cone, T. 0. Roggers, L. D. Hoy, and Dr. Kortright withdrew from the company, and returned to the States, making nine in all who had withdrawn since landing at Valdez. This greatly reduced our numbers, yet we pushed on as hopeful as ever that the future held a golden harvest in store for the plucky ones who deserved it.

One day while the sun was shining brightly upon the newly fallen snow over which I was pulling my loaded sled along up the trail, I was suddenly attacked with snow blindness. I was assisted back to the tent, and for two days suffered terribly from it, when I was able to resume work. It was no uncommon thing to see men taken back to their tents by their comrades, unable to see anything. All sorts of glasses were used as a protection, but in spite of any and all these, scores of men were stricken by it. The suffering from it is intense; the pain resembling that caused by the strongest fumes of freshly grated horseradish, and the only way we found any relief was by the liberal use of witchazel extract. We had happily provided an abundant supply of this in our stock of medicines, and it proved of inestimable service.That which afforded us the most protection was wooden goggles, the patterns for which were taken from those worn by Indians. These were whittled out of a piece of wood, fitting closely around the eye, and with no glasses at all but in the place of glasses were very small openings to see through, the inside being colored black. A projection like the visor of a cap extended over them, which was also colored black on the under side to shade the eye. The mechanical skill of the makers of these goggles was varied. Some of them were ungainly affairs, weighing half a pound or more, while others were thin and light, and beautifully carved.

At the fourth bench, where my camp was now located, there was a village of about one hundred tents scattered along the trail. For nearly a mile there was almost one continuous pile of goods. Men who had got their supplies farther up the trail were moving out, and going on up to the foot of the summit, where were perhaps one hundred and fifty tents more.

The higher we climbed upon the glacier, the more frequent and severe were the storms. Day after day we trudged along through the drifting snow, which was constantly filling the trail as we passed over it. It was not uncommon in the morning to find from six inches to a foot of fresh-fallen snow, through which we must break a trail but with hundreds of men in line, it was not so bad. Those who were in advance, when fatigued, would drop back, and others who had been having a comparatively easy time in the rear would take the lead and by so doing the trail would be opened without any great hardship to any one.

CHAPTER V

THE GREAT SNOW STORM

DAY after day we toiled on, and half our goods were at the foot of the summit, but we had consumed so much more time making the ascent than we had anticipated that our supply of wood was giving out, or getting short. We thought that we had enough to take us clear over; and we hoped that ten days would see us over the dreaded summit with all our stuff; but when we arose one morning, we found a blinding snowstorm raging, which, for severity, would equal any blizzard I ever experienced in the States. We had worked on many stormy days, and felt the necessity of improving every moment to the best possible advantage; but this was too severe, and thinking that it would cease in a few hours, we spent the day in our tents, keeping only fire enough to cook our food, and spending the balance of the time in our sleeping bags. Toward night we arose, and after supper went outside, and shoveled nearly a foot of snow from around our tent, but the storm showed no signs of abating. Next morning we awoke to find our tent nearly breaking under its weight of snow. We crawled out, and again shoveled the snow away, but the storm raged on. We went about setting our sleds up on end, and putting everything in a place where we could find them when the storm was over. We cooked and ate our breakfast, and again took to our sleeping bags, where we lay all day, and saw our tent gradually weighted down with another foot of snow. Again we went out, shoveled the snow back from the tent, cooked and ate another meal, and retired for the night, wondering what the end of all this would be.

I will not attempt to describe in detail the next four days, but night and day the storm raged on, and we could do nothing beyond shoveling our tent clear. When the wall of snow would become

"*When the wall of snow would become so high that we could not throw the snow over the top, we would climb out on top of it, and tramping the snow hard, would set our tent in a new place. Thus we kept on top.*"

Photograph by Neal Benedict from the Messer Collection courtesy of the Cook Inlet Historical Society.

so high that we could not throw the snow over the top, we would climb out on top of it, and tramping the snow hard, would set our tent in a new place. Thus we kept on top.

The smallness of our wood supply caused us to practice all sorts of economy in its use. When we had a fire, we would fill every dish we could spare with snow, and place it as near the fire as possible to melt, for this was our only way of obtaining a water supply. Frequently some man would come wallowing through the snow and inquire if we had any wood to sell, for which he was willing to pay almost any price. Of course we had none to spare, for we were compelled to see our own little woodpile almost entirely disappear, while the nearest wood was fully twenty miles away, and as impossible to get at as if it had been at the north pole.

Our short sleds, as I have before stated, were four feet long, and our long ones six feet. In the front ends of these sleds were steering poles six feet long, and when we set them up on end, we had not the slightest thought that they would be covered by the snow; but they had long since disappeared, and every day saw the pieces of the steering poles protruding above them growing less and less. On the morning of the sixth day of the storm, the handle's of the shorter sleds were entirely out of sight, and not more than one foot of the longer pole remained above the snow.

We had a few very small oil stoves and a small quantity of oil, which had served us a good turn, and we were offered prices for

"The smallness of our wood supply caused us to practice all sorts of economy in its use. When we had a fire, we would fill every dish we could spare with snow, and place it as near the fire as possible to melt, for this was our only way of obtaining a water supply."

Photograph from the *Wulff Collection* courtesy of the Valdez Museum and Historical Archives.

these which almost staggered us, but the selling of any kind of fuel was not to be considered for a moment. Snow-slides became so frequent that not three minutes would pass but what we could hear one somewhere in the mountains, and often they would be so great that the noise of them was like the rolling of heavy thunder.

One night the great mass of snow which had accumulated on the mountain side in the rear of the camp at the foot of the summit, gave way, and came crashing down the mountain. Nearly every one was in bed, little dreaming of the danger in store, when suddenly, with little warning, upward of twenty tents were buried under this great mass of snow from six to fourteen feet deep. Two of our tents, containing nine men, were among the number. No alarm had to be sounded, for everybody was awakened by the noise, and soon two hundred shovels in the hands of willing men were plying, to rescue the unfortunates who were underneath. Many men, barefooted, and clad only in their nightclothes—which were simply their underclothes,—set to work, and for hours labored with might and main to liberate their imprisoned tent-mates and comrades. Many were rescued just in time to save their lives.

My friend, Harry Sweet, of Hornellsville, N. Y., was the last of our party to be rescued, and says that he owes his life to the almost superhuman efforts of his tent-mates, who, as soon as they were liberated, sprang to shoveling, not thinking of their bare feet or scanty clothes, and shoveled with all their might till they saw that he was saved.

It may be of interest to the readers of these pages to give Mr. Sweet's story of his experiences upon this never-to-be-forgotten night. Relating them to me he said:

"I was the first to get into my sleeping bag that night. I had been in bed some time, and my companions had either retired, or were preparing to do so, when I heard a great roaring noise, which seemed to grow louder, and to come nearer. All at once it dawned upon me that a slide was coming toward the camp. I raised up in my sleeping bag, and shouted, 'Look out, boys, a snowslide is coming!' The words had hardly escaped my lips when it struck me, as I sat up in my sleeping bag, and pushed me over toward the center of the tent, where my tent-mate, Charles Priceler, was preparing for bed. The tent was crushed down on us in an instant, the side-pole coming across Priceler's feet, and as I was thrown over by the weight of the snow, my hand came in contact with one of his bare heels, which I clung to as well as I could. I do not know how long I lay in this position, but it was some time. I had no difficulty in breathing, but I could feel the great weight of snow settling down upon me. I could hear, now and then, the faint sound of shovels, as they occasionally clashed together, and I knew that men were digging for us. Soon I felt the foot moving upon which I had a grasp, and it quickly slipped away from me altogether. As soon as this occurred, it seemed as if the supply of breathing air was being shut off; and though I could hear more plainly the shovels of my rescuers, I began to feel that it was all up with me. I faintly heard these words: 'Let me have that shovel; I know just where he is.' I recognized it as the voice of Priceler, and knew that he had got out. I could feel my senses leaving me. My head began to whirl; I thought I was dying. I remember nothing more until I was being taken to a tent, and was put into a sleeping bag, and given a large drink of liquor. It seemed to warm and revive me, and I soon felt quite comfortable.

"During the night some one came to the tent, and yelled, 'Look out! another slide is coming.' I was out of that sleeping bag in a jerk, and went dashing after my companions as they swept out of the tent. We started across the glacier, floundering in snow nearly to our waists, trying to get out of harm's way. The alarm proved to be a false one, and we waded back to the tent more leisurely. On my way back I met a woman, and having only my underclothing on, in which I flew during the scare, she said to me, 'My dear boy, why don't you get on some more clothes ?' I answered, 'My good woman, I haven't any.' On my arrival at the tent I borrowed some to wear until I could get mine out from under the snow, where they had been left. Everybody remained up the balance of that night, and several trails were made for some distance out on the glacier to enable us more readily to escape, in case of another slide."

After Mr. Sweet had related to me this experience, he told me of two other very narrow escapes which he had had while railroading, being mixed up in two wrecks, and now that he had survived the snow, wondered what would come next. We all hoped that this would be the last.

Two poor fellows were taken from under this great snowslide dead. Being Odd Fellows, their remains were taken in charge by members of that order; and as soon as the trail was opened so it was possible, they were wrapped in their blankets, placed upon sleds, and taken over the summit, down to the timber, where graves were dug, and they were buried.

At this same time another slide occurred at the second bench of the glacier, in which several men were injured, and seven mules were killed. It was supposed at the time that no men were killed, but there came a report late in the summer, that the melting of the snow had uncovered the remains of three men who had perished under it. I can not vouch for the truth of this report, but I do know that whole tents full of men might have remained buried under that slide at the summit and not have been missed; for people generally were strangers to each other, and during the storm few had stirred out to ascertain who their neighbors were.

The day following the snowslide was spent in moving our tents farther out upon the glacier, to a place of safety. Then began the breaking of new trails, which, with wind blowing and snow drifting, was a hard and slow business. A dozen men went over

The avalanche of April 30, 1898. "Many men were out with long poles, probing the snow in search of lost goods. Frequently a man's stick would strike a pile of goods, and upon shoveling them out, he would find that they belonged to some one else; and while he was doing this, perhaps some other man might have discovered his." Crary Collection, B62.1.2026. The Anchorage Museum of History and Art.

the trail with snowshoes, packing as well as they could. Following them were as many men as could be got out for that purpose, tramping a trail wide enough for a good road, and following these came lightly loaded sleds. In this way a fairly good trail was opened up from the fourth bench to the summit, in two days. It was the aim of the trail-makers to get the new trail as near as possible to the line of the old one, for scattered along that were hundreds of caches of goods which now had no mark by which to locate them; for everything that might have been left to designate their location was now many feet under snow. Many men were out with long poles, probing the snow in search of lost goods. Frequently a man's stick would strike a pile of goods, and upon shoveling them out, he would find that they belonged to some one else; and while he was doing this, perhaps some other man might have discovered his. In this way the greater portion of goods along the line was discovered. But occasionally a man would probe for his goods many days without finding them, and even then perhaps he would be compelled to abandon his search; nor is this strange when it is remembered that this single fall of snow measured fully eleven feet. Besides this, thousands of dollars worth of goods were carried away by the great "slide," beyond recovery, thus leaving destitute many poor fellows who had labored hard, almost night and day, to get them up thus far.

As soon as the trail was open, the two tents which had been at the fourth bench, were moved to the foot of the summit, so our five tents were now together, and the goods remaining below were hurried along up the trail as fast as possible. While we were doing this, there came a report from the interior to the effect that those who were in advance had reached the valley beyond the glacier, and following it down, had found it only a barren, surrounded on all sides with high mountains, with no possible chance to go farther. This practically put a stop to all work on the trails until the truth or falsity of the report could be ascertained.

Various plans were proposed, but the one which promised the quickest solution was to send a man back down the glacier to "Fox Island," to see the Indian woman, whom I have mentioned as having her feet frozen while passing over this very place, and ascertain of her if we were on the right trail to Copper River. A man in the crowd was found who knew her, and he consented to go. He must necessarily take a rowboat from Valdez, and so a collection was taken to hire two men to accompany him, and he immediately started on his errand. In two days he returned with the welcome news that we were on the right trail. This report had undoubtedly been started by some parties in advance who hoped to turn back some of the large crowd which they knew was following them. Of course it would not be difficult to imagine their object; for if gold should be found down the valley, with so many people after it, there would be a most interesting scramble for claims.

CHAPTER VI

OUT OF WOOD ON THE SUMMIT OF THE GLACIER

OUR small supply of wood became exhausted, and one by one the few boards which we had brought along to serve as a floor in our tents, to keep our sleeping bags off the snow, were used up. The board we had used as a table was next to go, and then we had nothing left to cut. Something must be done. So about three o'clock one morning Wesley Jaynes and I set out to find wood. Taking our sleds, we decided to take a load of goods down the opposite side, toward the Copper River Valley, and bring back wood. We climbed to the summit, where we had a cache of flour, loaded three hundred pounds on each sled, and began the descent. We drew our sleds the first half mile with but little exertion. Coming then to a place where the sleds would run themselves, we climbed on, and ran about a mile, riding the entire distance. Here we came to a steep place. The day before had been warm enough to thaw, and during the night a crust had frozen sufficiently strong to hold our loads with our own weight added. For half a mile the glacier was so steep that it was with difficulty that we could hold our loads back. When near the bottom of this steep bench, we could look for miles down the glacier, over the smooth crust, with nothing to impede our progress. I went to the left of the trail some distance, and mounting the load, let it go. The first half mile I rode faster than I cared to, perched on a load of flour; but meeting no obstacles I rode on for three miles, thinking of what a "snap" we were going to have when we should get all our goods on the summit, and expecting my friend Jaynes was following me. But upon glancing back, he was nowhere to be seen. I stopped my sled, and began wondering what had become of him. Taking a survey of the situation, I soon saw him away off to one side of the glacier, walking along without any sled. I was wondering what had become of it, when it occurred to me that it had run away. In a little time I saw him coming toward me, and he had his sled. I waited for him, and he related to me his experience. He had mounted his sled farther up the steep hill than I had, and had gone down on the right side of the trail, where the surface was not as smooth; and upon reaching the foot of the steep hill his sled had acquired such a velocity that in going over a rough spot he was thrown off; and the loaded sled shot on down the glacier by itself. He followed after it for two miles, and when he came up to it, found that it had stopped just on the brink of a great chasm in the glacier, hundreds of feet in depth, whence, had it made the leap, it would never have been recovered.

The next four miles there was just enough descent for the sleds to run themselves, and we had, another delightful ride over the smooth crust. The descent then became too great to ride, and we walked the next mile and a half, holding our sleds back. Here we reached the last half mile of the glacier, which was at an angle of forty-five degrees. Wrapping a piece of large rope around one of the sled runners to assist in holding back, we easily made the descent to the foot of the glacier.

The valley leading down from the foot of the glacier has an average width of one mile, down which flows one branch of what forms the headwaters of the Klutina River, but which was now frozen solid, and covered with three feet of snow from mountain to mountain. This was also crusted sufficiently to hold our weight, and taking our loaded sleds, we started down the valley toward the nearest timber, which was four miles farther on.

Here we found a little village of about twenty-five tents. These men were in advance, and had got their goods all over the glacier, and were pushing on down to Lake Klutina, which, they informed us, was seventeen miles distant from the camp.

Unloading our flour, we ate our dinner, which the crisp, cold air had frozen, after which we began to look about for some dry wood. Noticing some fine dry trees about a quarter of a mile distant, we started across the valley toward them. The warm sun had for two hours been shining down into the valley and had softened the crust so it would not bear our weight; and about every other step we would break through, and go down nearly to our waists in soft snow.

At last we reached the timber, and cutting twelve sticks, five feet long and with an average size of four inches through, we loaded six upon each sled, and started for the foot of the glacier, where we arrived about 3 P. M. We found the ascent very hard work. The sun had softened the snow in the trail so that we sank into it six inches at every step. Reaching the top of the first bench, we sat down to rest. Hearing a slight noise up the mountain, we glanced upward, and fully five thousand feet above us, saw the beginning of a snowslide. Knowing that we were in a safe position, we sat and watched it. At first it was small, and the noise of it could scarcely be heard; but as it came down, it increased in width and force, until it was sweeping everything before it.

About halfway down the mountain side there was a perpendicular wall of rock, perhaps a thousand feet high. When the great mass of snow reached the brink, it swept over, and when it struck the rocks below, we could distinctly feel the great glacier upon which we were sitting tremble from the shock. Then it started again, on another downward sweep, rushing along the steep mountain side until near its foot, where was an incline, down which it seemed to run like molten lava. And I could then see how so many people had been buried under these great avalanches.

Starting on again, we pursued our upward way, occasionally stopping to rest, until nine o'clock at night, when we had reached a point two miles from the summit. Here we became so thoroughly exhausted that we could take our loads no farther, and leaving them beside the trail went on, reaching camp more dead than alive. Next morning we returned, and brought in our loads. We could have sold these two little loads of wood before reaching camp for forty dollars, so great was the wood famine. These two small loads, when divided among five camps, gave only a small amount to each tent, and we practiced the most rigid economy in its use.

Occasionally some men would get their goods all on the summit, and then move their camps on to the timber; and we would watch eagerly for these chances to gather up tent stakes, or any small pieces of wood which they might leave behind. Dry wood generally sold for twenty-five cents a pound, and often could not be had at that price.

At last we had all our goods at the foot of the summit, and the next move was to get them upon the top. We established relays of men up the summit, and one man would take a sack of flour and carry it a hundred yards when another man would take it, carrying it another hundred; then a third, and fourth, and so on. In this way one hundred sacks were moved to the top, when this plan was abandoned, and the rope and pulley again brought into use. We stretched one thousand feet of rope, and sent fifteen men to the top, who seizing hold of one end of the rope, came down, thus drawing up a sled load of ten to twelve hundred pounds on the other end.

We would haul all our goods up thus far, plant our stake another thousand feet up, and repeat the operation. It took five pulls of this sort to get our goods from the foot of the summit, or the last bench, to the top of the glacier.

All this had to be done in the severest weather, and frostbites were of common occurrence. But that which gave us the most trouble was the snow blindness. Often a man would be taken blind while on the trail, and it was no uncommon thing to see one man leading another back to his tent, where he would suffer for days, unable to see anything.

One very remarkable incident occurred, which I will here mention. One miner had four magnificent dogs buried under the snowslide. When it was thought that all the men had been rescued, he set about rescuing his dogs, if possible. After hours of hard work, he came upon one, which was dead. He kept on till he reached the second, which was also dead, and so on to the third. Finding these three dead, and being nearly exhausted himself, he gave up the search for the fourth dog, and went about his other work. Eight days after as some men were probing in the snow for lost goods, they heard the faint bark of a dog coming from beneath the snow. Securing shovels they dug down, and rescued the dog, which in a few days was apparently as well as ever.

CHAPTER VII

DESCENDING THE GLACIER

AT last our goods were all on top of the glacier, and the time had come when we could move our camp over to timber. I shall never forget our last two days' work. We worked in a blinding snowstorm, the wind blowing almost a gale, while the air was keenly cold. For twenty-six hours we toiled on without rest, over a bad trail, but finally reached our destination at the timber line, so tired and worn out that the boys scarcely waited for anything to eat, but threw themselves down upon their blankets, and were soon fast asleep.

By this time the sun had completely ruined the trail leading from the foot of the glacier down to where we were camped. During the middle of the day men would go in to their knees in slush, if they attempted to travel it. But the nights were so cold that a crust would form on the snow sufficiently hard to bear our weights, so we would leave our camp before midnight, go to the summit, and bring down a load of five or six hundred pounds, unloading part of it at the foot, and bringing the balance on the crust to camp. Our second trip to the summit after goods was an unusually severe one. We left camp about 11 P. M., and started out up the glacier. But when about half way up, we encountered a snowstorm so severe that we were compelled to return, and wait a more favorable time. The next night was bright and clear and we considered it an ideal time to start for the summit. The night air was crisp and frosty, and when half way up, the wind began to blow, and the snow, which was almost constantly falling on the summit, began to come down to us, and it was not long before we were in a terrible blizzard.

Camp at first timber over glacier.
Photograph reproduced from first edition.

We donned our oil-skins, to protect ourselves as far as possible from the storm and cold. The trail was entirely obliterated, and we were obliged to put a man in advance with a long probe-stick, to feel out the trail. This made our progress extremely slow, and for hours we pushed on into the teeth of the storm, which was constantly growing worse as we neared the summit. We met several men feeling their way down with loaded sleds, as we were feeling ours up. They told us of the severity of the storm farther up, and said we had yet two miles to go, and advised us to return with them; but we had struggled so hard to come thus far that we decided to push on and reach our goods if possible. So we passed the men, and in a moment there were no signs visible of the trail over which they had come.

Up and up we went, wondering how much farther it could be, when at last our cache of goods was reached, almost buried out of sight in snow. We tarried here only long enough to eat our lunch, which by this time had become frozen in our pockets, having all the time to stamp our feet and whip our arms about our bodies to keep up circulation, and prevent our freezing.

As soon as possible we loaded our sleds with five hundred pounds each, and began our return; but it was slow work, for our leading man had to thrust his sharpened pole down into the snow at almost every step to keep from losing our way, and to lose the trail in a storm like this meant hours of floundering about in the drifting snow, and then almost certain death from the biting frost. But with all our caution, we often found our sleds bottom side up in the deep snow beside the trail. This would call a halt of the entire party, for we dare not become separated. Those nearest to the unfortunate would assist him to right his load, and then we would proceed. But who that has not experienced it can imagine the relief when, as we neared the foot of the glacier, we came out into the brightest sunshine, and were able to complete our trip with comparative ease?

By the time our goods were all got down off the glacier, the snow in the valley had gone, and left us with three tons of goods still up at the foot of the ice. Here our horse served us a good turn. We made for him a sort of Indian drag sled upon which he could haul three hundred pounds at a load, and make two trips a day. While the men were at other work, the horse was kept busy until all our goods were at our camp four miles from the glacier.

The snow was gradually disappearing along the valley and the foothills, and the stream near our camp, leading to Lake Klutina, was slowly rising, and we were convinced that it would not be long until we could boat our goods down. But the descent was such that the current must necessarily be very swift, and we knew that to get a boat back up stream after having taken a load of goods down, would be a difficult task. So we decided to build as many boats, and with as great carrying capacity, as possible.

Near where our camps were located there had been trees sufficiently large to make good lumber, but these had all been cut by men in advance of us, and the lumber from them was fast being made into boats. Three miles below our camp was a fine piece of timber, and there we decided to establish our shipyard. We could not move our camp to this spot, for as yet we had no boats. So it was only left to us to walk three miles every night and morning to and from our work. In going this distance we were obliged to cross the river five times. At first this was no great task, as the water was low; but every day we could see that it was gradually rising, so that at the last it was accompanied with great danger, for if a man should lose his footing in the swift current, he could not regain it, and must surely be carried down.

Manufacturing lumber for these boats was necessarily very slow and hard work; for it must be borne in mind that we had no sawmills along with us. Our only saws were long handsaws, or long slitting saws, and our only power was muscle power. The method of making logs into boards is novel to those unaccustomed to the methods of frontier life. Sound, straight trees as large as can be found are selected, which in this country are usually about fourteen inches in diameter, and these are felled and cut into logs as long as one desires the boat to be. Then two large sawhorses are built about six feet high, which are placed as near the logs as possible; then if you have a small company, your companions are invited to a "rolling-bee," and a log is placed on the two horses.

The log is then peeled and leveled up, after which, by the use of a plumb-bob, a perpendicular line is made upon each end to mark the center of the log. Measuring each way the thickness of the lumber desired, other marks are made on top and bottom; then lines are struck along both top and bottom from each of these points. Thus the whole log is laid out into lumber of the desired thickness ready for the saw, and all that is required to manufacture good lumber is a good saw, experienced hands, and plenty of elbow grease. As fast as the lumber was sawed, it was carried on our backs out of the woods to our shipyard, where it was piled up in a position to dry.

Near where our camps were located were the graves of the two men who were killed upon the glacier in the great snowslide. The Odd Fellows had graded up the little plot of ground containing their graves, and bordered it with round stones taken from the bed of the creek. On the 30th of May, our national Memorial day, the miners who were encamped at this place went out and gathered a profusion of wild flowers, and decorated the graves of these two companions in hardship, who had so suddenly fallen and simple as these ceremonies may seem to the reader, they were nevertheless sad to us.

CHAPTER VIII

OUR FIRST BEAR HUNT

A FEW days after this, Captain Moyes, Wesley Jaynes, and I took a day off, and went hunting. We had heard of some bears being seen upon the mountains, so we were up early, and had our first experience in real mountain climbing. We were armed with our trusty old Winchesters. The snow remained only in spots, and for half a mile we climbed, holding on to bush and rock to keep from sliding back. When at this height, we came upon the tracks of a large bear, and also of some mountain goats. We also found ourselves confronted by a wall of rock fully one thousand feet high. There was only one spot where it was possible to make the ascent, and up through this the bear had gone. When there was any difficult undertaking ahead, the men on the trail would say, "It's a tough proposition." So this seemed to us; yet we started up, but more than once regretted having done so.

Now and then we would halt, and think to go back; but one glance down would convince us that the ascent was less dangerous than the descent, and we climbed on up. Finally we reached what we supposed was the top, but we found, to our dismay, another thousand feet of rocky wall. We followed the bear's track along this ledge until we came to a small ravine, up which he had gone. This was made more easy of ascent by the hard snow which remained in it. Up this ravine we climbed, scarcely daring to look back, until we passed over the summit, and, with a sigh of relief, rested upon level ground fully six thousand feet above the valley we had left in the morning. After a brief rest, we started on the bear's track again, and soon came across some fresh moose tracks. This caused us to leave the bear, and go moose hunting. Soon we came upon so many tracks that it was impossible to follow any of them, they were so mixed up, and we were compelled to give up the chase.

By this time it was noon, and we sat down to eat our lunch upon a ledge overlooking the valley so many thousand feet below, and it was a sight to waken in one the desire for an artist's brush and an artist's skill. The picture was indeed one of rare beauty. Two miles below was a village of a hundred tents, among which were our own. Four miles farther on were at least two hundred tents, in plain view. The little stream was lined with busy men, floating their goods down to the lake, and scores of others were building boats. Farther down, other camps were seen, and nineteen miles away the lake was visible. If we had cared to look up the valley, there would have loomed before us the great glacier, upon which we had labored so hard and long; but we had seen enough of that, and turned our eyes to pictures that were new. The day passed without our catching sight of any large game except one lynx, after which I sent a couple of shots as he passed out of sight around the rocks. We saw scores of woodchucks, which, by the way, are nearly white in this country. Ground squirrels were almost always in sight. We also saw several ptarmigans, a few of which we shot. They are similar to the partridge of the East, only they are pure white in winter and brown in summer. At this time, June 4, they had about equal parts of their old and new coats, making them speckled. We found a place in the mountain more easy of descent than the route by which we had climbed, and reached our camp without mishap, but fully persuaded to do the balance of our hunting on the lowlands.

At this time of year there is scarcely any night here. The twilight begins about 11 P. M., and by 1 A. M. it is daylight. Even at midnight one can easily read a newspaper inside the tent, without artificial light.

While our company was camped at this place, our president, D. T. Murphy, of Stamford, Conn., and one of the founders of the company, left us for the States, giving as his reason for so doing, the dangerous illness of some members of his family. Shortly after this, Captain Emanuel Moyes, Charles Butts, and William Williams made application to withdraw from us also. They were given a certain portion of the goods, and from this time on they continued their search for gold independent of the company. Our numbers were diminishing, but still we pushed on, as confident of success as when we left our homes fully five months before.

CHAPTER IX

BUILDING BOATS ON THE KLUTINA

AS soon as we had a sufficient amount of lumber sawed, we began the building of our boats. The work of boat building was much more difficult than it would have been could we have stepped into a planing mill and selected what we needed, already sized in width and thickness. But here everything wanted must be taken from the tree, and worked out, and this, too, with but a limited selection of tools.

Our boats were built twenty feet long, and had a carrying capacity of about three thousand five hundred pounds each. Walking to and from our work was a ruinous waste of time, and as soon as one boat was completed, we began moving our camp down to our work. The river by this time had become a roaring torrent, and almost every day we would see goods floating by on its surface, from somebody's boat which had been capsized in the river above. Many men had been building boats who were entirely ignorant of how they should be constructed, and in many cases they were entirely destitute of the elements most needed to navigate such a boisterous stream as this in safety. Into such boats as these they would load their goods. All would go well until the boat struck a rock, which it was sure to do before going far, when it would go to pieces, and the goods which it had carried would come floating down the river. Often men became so excited at the prospect of losing their goods that they would rush into the swift, icy water, and endanger their lives to save a sack of flour or some other articles of their stores.

I once saw a man coming down the river with a boat load of provisions, purchased with money for which he had mortgaged his home. A few months before, he had left his dear ones, so full of

The work of boat building was much more difficult than it would have been could we have stepped into a planing mill and selected what we needed, already sized in width and thickness. But here everything wanted must be taken from the tree, and worked out, and this, too, with but a limited selection of tools. Photo from the first edition.

hope that he could soon return with sufficient means to provide for their every want, and after months of toil and exposure had, as he thought, nearly reached the country where his fondest hopes were to be realized. But suddenly his boat, which contained all his earthly possessions, while rushing through the swollen waters, struck one of the countless hidden or protruding rocks, and quickly went to pieces, and his goods were scattered among the driftwood along the banks or at the bottom of the merciless water. And standing there upon the bank, his goods all gone, his hopes all blighted, the great tears rolled down his cheeks as he thought that he must now find his way out of this country, and go back to his loved ones empty-handed. It was a sight to move the heart of any one having any sympathy left in his nature. But it was among the common incidents of a mining country, or a country where such an army of fortune-hunters were searching for gold.

One by one our camps were moved to a spot near where we were building our boats, which enabled us to push our work much faster. We had brought with us a sufficient quantity of oakum with which to calk our boats, but only half enough pitch. So men were sent out to gather an added supply, of which there was an abundance on the spruce trees growing in the nearby forests.

At last our boats were completed and launched, and we had the satisfaction of seeing a fleet of six fine scows drawn up in line in front of our camps.

We had heard rumors of the struggle pending between the United States and Spain, and it was suggested that, as soon as our goods were successfully landed at Copper River, we should tender the use of our fleet of boats to the United States (Uncle Sam), for with such a formidable fleet as ours at his command, little doubt could be entertained as to the result of the conflict.

There must have been at least a thousand boats built between the foot of the glacier and Lake Klutina; and there was almost as great a variety as there were boats. Some men spent many days in building crude log rafts, upon which they expected to load their goods, and float them down to Copper River. Even after accidents with boats had become frequent, men persisted in building rafts, and risking their goods and even their lives upon them in these madly rushing waters. But I do not recall a single instance where goods were safely conveyed on such crafts down this turbid stream for any considerable distance.

Often some expert mechanic who had brought along a good variety of tools would turn out a boat fine enough to grace the waters of any aristocratic summer resort, but often these finely constructed boats would be the first to go to pieces when put to the test of carrying loads down this swift mountain stream.

As soon as the snow was sufficiently gone, some of the men began to prospect for gold in the valley of the Klutina and its tributaries, but the larger portion of them pushed on toward Copper River. Hunting parties were also out, penetrating the forests, ascending the foothills, and climbing the mountains in their search for game.

At night camp fires were visible in almost any direction one could look. The moss and brush by this time had become very dry, and as a result of the carelessness of campers in leaving their fires, forest fires began to rage along the valleys. Even the green tops of the spruce trees would burn like tinder, and the flames would shoot upward into the sky for a hundred feet above their tops. It was a beautiful sight, but it seemed too bad to see hundreds of acres of finest spruce timber thus destroyed.

Very naturally this was a source of grievous displeasure to the Indians to see the forests destroyed through which they had roamed from year to year, and which afforded shelter to the game without which it would be difficult for them to subsist. But they seemed to have no thought of retaliation, and always treated us in a manner which bespoke friendship for the white man.They would sometimes visit our camps, and when successful in their chase, had no hesitation in bringing us a choice steak of moose, caribou, or mountain sheep, as the fruit of their hunt made possible.

With six boats at our command it was but the work of a few days to get all our goods down to the present location of our camp. This was done without any serious mishap or loss.

One day a member of the company who was in charge of a boat which had made several successful trips, was, in company with others, bringing down loads. As they approached a dangerous spot in the stream, some one on one of the other boats cautioned him to be careful, but with an air which bespoke his ability to pass safely all the bad places, he said, "Oh, I know every rock in the channel." Scarcely were the words uttered when his boat struck a rock, and immediately filled with water, but remained fast against the obstruction. The larger portion of the goods remained in, but some of them floated out and down the stream. The channel happened to be narrow at this point, and a part of the crew jumped out and got to shore, and running down below to a point where the current brought the floating articles near them, they fished out the greater part as it passed.

CHAPTER X

OUR FIRST PROSPECTING EXPERIENCE

As soon as our goods were all down, our entire company took a whole week off, and went prospecting. We divided into small parties, and went out in every direction, visited all the tributaries of the headwaters of the Klutina, digging down as far as the water would allow us, and washing out pans of the dirt at different intervals, as we dug down.

We found some gold on almost every stream; from forty to seventy colors could be counted in every pan washed out. But it was of the kind known as "flour gold," so fine that it could not be obtained in paying quantities.

The week of prospecting only brought this kind of discoveries, but it served to keep us hopeful that when we penetrated farther into the interior, we should find some richer strikes. So we began preparations to move down seventeen miles to Lake Klutina.

The fourth of July was near at hand, and the boys were planning for a grand celebration. Among the three hundred people camped at this place, there were some very excellent singers, so it was arranged that upon the evening before the Fourth a concert should be given at one of the largest tents. This proved a very enjoyable affair, and lasted until nearly midnight.

As soon as twelve o'clock came, it seemed that every man who was the possessor of a gun of any kind began to celebrate by wasting cartridges in the air, and for five minutes there was such a roar of firearms as I feel safe in saying was never heard in this section before. Many remained up all night, and passed the time in keeping every one else awake also. The hundred guns planned for a sunrise salute was so enlarged upon that it came much nearer to one thousand.

At ten o'clock a parade was given, with all the miners in line.

"The fourth of July was near at hand, and the boys were planning for a grand celebration. Among the three hundred people camped at this place, there were some very excellent singers, so it was arranged that upon the evening before the Fourth a concert should be given at one of the largest tents. This proved a very enjoyable affair, and lasted until nearly midnight."

Joseph Bourke photo, courtesy of the Wulff Collection, Valdez Museum and Historical Archives.

This differed in many respects from any parade I had ever witnessed before—there were no spectators. The men marched and countermarched back and forth along the trail which led through the village of tents, each carrying some article used in camp life. At one o'clock a game of baseball was played upon a large level sandbar, a short distance from the river. Following the ball game, came sack races, foot races, and various other games. One of the most novel features of the day was given by a Minnesota man, in riding a log down the turbid waters of the Klutina River.This proved to be the most exciting exhibition of the day, for the man was not always on top of the log; but he exhibited great skill in the art of log riding, and it became evident that it was not his first attempt. He did this feat standing straight up on his log, with simply a pole in his hands to use in balancing or guiding his treacherous craft.

That same day, after the celebration was over, we decided to move on sixteen miles to the lake. None of us had been any farther down than where we then were; the stream was unknown to us, except from hearsay, and it was reported dangerous. We had seen many men coming back who had lost all their goods before reaching the lake, and it had caused something of a dread to come over us all, yet we must attempt it. So all the boats were brought up, and loaded with our camp outfits, and as much goods as we considered safe to take, and started out down the river, allowing

"We were just priding ourselves on our good fortune in
escaping the shoals, when we
heard our boat grinding on
the bottom, and presently it
came to a standstill. Now
came our turn to get out into
the water, and after a half
hour's hard work we had the
satisfaction of seeing her
float again."

*Neal Benedict photo. Messer
Collection courtesy of the Cook
Inlet Historical Society.*

about one quarter of a mile between boats. The water was high and
muddy, and it was impossible to tell where the deepest channels
were. My boat was the last to start, and we had not proceeded far
when we saw one of the others stuck fast on a sand bar in the
middle of the stream. The men were out in the water trying to get
her off the bar. We asked, when near enough, if they needed any
help, but receiving a negative answer we shot past them, and were
soon out of sight around a bend in the river.

We were just priding ourselves on our good fortune in escaping the shoals, when we heard our boat grinding on the bottom,
and presently it came to a standstill. Now came our turn to get out
into the water, and after a half hour's hard work we had the satisfaction of seeing her float again. We climbed into her as she swung
off into deep water, and away we went down the stream, now
pulling hard to avoid a rock on this side, and in a moment pulling
as hard to keep clear of flood-wood on the other, around which the
water would boil in maddening fury. Often we would see on shore a
stake bearing a red cloth, as a danger signal. Keeping a sharp lookout,
we would soon discover some sunken tree, or other obstruction
upon which we might easily have been wrecked, had we not been
forewarned.

Barring a few slight mishaps, our boats all reached the lake
unharmed. We pitched our tents close to the river's bank on an island of about twenty acres. It was indeed a beautiful spot, and an

*"We climbed into her . . .
and away we went down
the stream, now pulling
hard to avoid a rock on
this side, and in a moment
pulling as hard to keep
clear of flood-wood on the
other, around which the
water would boil in mad—
dening fury."*

Joseph Bourke photo, courtesy
of the Wulff Collection, Valdez
Museum and Historical
Archives.

ideal place for a camp. We could look three miles across to the opposite side of the valley, and see another stream, which came
down a narrow divide and emptied into the lake, and was as large
as the one upon which we had come.

Along down the opposite side of the lake the foothills extended back for eight miles to a range of mountains fully six thousand feet high. The lake was twenty-eight miles long, but only the
upper half was visible from our camp. The upper half was five
miles in width, and the lower half about two.

At the rear of our camp, an arm of the lake one-half mile wide
extended into the mainland one mile. Beyond this were the foothills, which ran back two miles to another high mountain range.
Upon the little island where we were camped there were only
about twenty-five tents besides our own, the larger camp being at
the opposite end of the lake.

The next day we fastened three boats together, and twelve
men pulled them back up the river. This was indeed hard work,
for the current was swift, and in many places the shore was ob-
structed with brush so thick that it was almost impossible to get
along at all. But we kept on, and after eighteen hours of hard pulling reached our cache of goods, too weary almost to rest. However, we only stopped long enough to prepare and eat our meal,
when we loaded our boats, and again set out for the lake. We got
some rest, but no sleep, on our down trip; and everything going

well, we reached the lake at six o'clock, having been absent from camp twenty-two hours.

The Eastern reader may wonder how we could navigate such a turbulent stream in the night, but he needs only to be reminded that there are no nights here at all at this season of year.

For the last month the mosquitoes had been a source of great annoyance to us, but at this time it seemed that it had reached a climax. They came by millions, and gave us no rest night or day. During the summer months we were obliged to go about at all times with our heads incased in mosquito netting, and long-wristed mittens on our hands.

Occasionally some one would return from the Copper River and tell us that we had no mosquitoes at all, but just to wait till we got there, and we would know what mosquitoes were. We wondered how they could be more plentiful anywhere than where we were, for it was almost impossible to remove our masks long enough to eat. These pests are not so large here as in the States, but what they lack in size they more than make up in numbers and viciousness.

We were obliged to make several trips up the river, before all our goods were brought down, and we were not sorry when the last boat load reached the lake. Almost every day somebody's boat would be capsized along this part of the river, and the goods scattered and lost. This was due, in a great measure, to a lack of skill in managing the boats, for it required skillful handling, together with much hard labor at the oars, to bring a loaded boat down in safety. But every day men who had never handled an oar would start down with an overloaded boat, and in nearly every case these were the men who had the accidents, and met the losses of their goods.

CHAPTER XI

POOR CHARLES KELLEY,

AND OTHER UNKNOWN UNFORTUNATES

WHILE walking to and from our work when we were building our boats, I remember passing a man at work on a raft. Many days were consumed upon this rude craft, only to find in the end that it could not be trusted to convey his stuff down the river to the lake. So one day he informed us that he was going to build him a small boat. Moving our camps below where he worked, we saw him no more. The days lengthened into weeks, and the little man and his boat had been forgotten. We had camped several days at the lake, when one day, as several of us were out in front of our camp beside the river, we chanced to glance along up the river, and saw a small boat coming down, bottom side up. Taking a boat hook, we waded out some distance in the stream, and pulled it ashore. Tipping it over, we found half a boat load of goods under it. These were placed upon the bank, and the boat made secure. It was a poor excuse for a boat, and resembled one as much as an ordinary dry-goods box. It was built entirely of slabs and short pieces of boards, and at its best must have leaked badly. We inquired through the camp if any one knew the owner of the boat, and finally ascertained that it belonged to the little man before mentioned, whose name was Kelley.

A search was instituted for his body, but it was unsuccessful. A week later a body, supposed to have been his, was seen to go through the rapids, thirty-five miles below. A miners' meeting was called, and a committee was appointed to take charge of his effects found in the boat, and to ascertain, if possible, the address of his relatives, and if so, to dispose of the goods to the best possible advantage, and forward the proceeds to them.

All the goods were looked through without gaining a clue to his identity, until at last we found, tucked away in a little bag containing needles, thread, and buttons, as if placed there by his wife or daughter, a small memoranda, and upon the fly-leaf of this little book, written in a feminine hand, these words: "My name is Charles Kelley, was born in 1844; height, 5 feet 5 inches; weight, 165 pounds; wear 7+ hat; No. 9 shoes, 16-inch collar, 16 shirt; my address is 159 Huntington Avenue, Providence, R. I. In case of accident, notify Jane Kelley, same address." This explained the whole matter. And as soon as possible the goods were advertised to be sold at auction the next day; but they were in such a damaged condition from being so long in the water that they brought the small amount of $17.

A letter was written to Mrs. Kelley, giving an account of his death as far as we knew it, and enclosing the money, with the names and addresses of several of the men who had known him on the trail. This was done in case more proof of his death should ever be needed.

It was afterward learned that he was seen passing a camp ten miles up the river only a few hours before we had fished out his boat, by some men who knew him, and that he remarked to them in passing that he had come near being wrecked a little way up the stream. Four miles farther down were several bad places, where many better boats than his had been capsized, and it was evidently at one of these that Charles Kelley found a watery grave.

Across from our camp, on the opposite side of the river, standing alone by themselves, were a tent and a cache of goods, which evidently belonged to three men. Everything inside the tent, and also without, indicated that its occupants had left, intending to be absent but a few days; but at the time of our leaving the place they had been absent ten weeks, and nothing about the premises had been in the least disturbed. It was our belief that these men had been drowned while crossing some mountain stream, or had met their fate in some other way. Whether this was true or not we never learned.

CHAPTER XII

CATCHING SALMON, AND RUNNING FOR A CLAIM

WHEN we arrived at the lake, the salmon had been coming up the river for some time, and as a result the lake was full of them, and it was no trouble to catch as many of these red beauties as we desired. They would not bite a hook, but there was no need to fish for them with hooks. We manufactured hooks out of spikes, by bending them up and sharpening the points, and fastened these to the ends of poles from eight to twelve feet in length. Armed with these, we would stand upon the bank where some clear stream emptied into the river; and the fish swimming up into the clear water by the score, we could then hook them out in large numbers, weighing from eight to twelve pounds each. These fish were excellent eating, and were greatly relished after eating bacon for so long. During the summer we used large quantities of fish, and it proved a great saving to our stock. We also had two spears, but they proved too small for such large fish as these. We had no difficulty in striking them, but to land them into the boat was quite another thing. We tried all manner of schemes for catching these fish, but aside from the seine and the gill net, our manufactured hooks proved the most effectual.

The Indians' device for catching them is an oblong willow basket of good size fastened to a long handle. Standing on a plank, or some other object, a few feet out over the water, they place the basket up the stream as far as they can reach, and let the current carry it back down the length of the handle, when it is lifted out, and the operation repeated; the fish, always going up stream during the early summer, run into the basket, and are caught. Often a score of these fish are caught in this way in a few hours. They then dress them, and put them on poles to dry. These fish, which by the way are usually caught and dried by the squaws, form a large and important part of the Indians' supply of food for winter.

"The Indians' device for catching them is an oblong willow basket of good size fastened to a long handle. Standing on a plank, or some other object, a few feet out over the water, they place the basket up the stream as far as they can reach, and let the current carry it back down the length of the handle, when it is lifted out, and the operation repeated; the fish, always going up stream during the early summer, run into the basket, and are caught." Photo reproduced from the first edition.

Nearly every company on the trail erected a place for smoking and drying salmon, and tons of them were thus put up for winter use. But the white men failed in curing their catch, though the Indians had practiced it for no one knows how long; and before the summer was past; car loads of them were spoiled and dumped into the river

We had been camped by the lake but a few weeks when it became known that a company of men were operating a sluice-box in a certain gulch some six miles from our camp, and it was reported that they were taking out gold in paying quantities. The clean-up, however, was always made when no one was present but some one directly interested in the reported find, and it was impossible to ascertain the amount of gold being taken out.

Another rumor came floating around one day to the effect that these men were finding gold in large quantities, and that the gulch was being rapidly staked out into claims. This caused a stampede from the camps. I must confess that we did not have so much confidence in the reports as did many others; but it wouldn't do to take any chances of getting left, so some of our fastest men, with very light packs, joined in the rush, and made the race for the gulch, while others, with provisions and blankets and tents, came on at a slower pace.

The country over which the stampede was made was covered with fallen timber, brush, and rocks, which made it most difficult to get along at all. But our "runners" reached the spot in time to stake out three claims. We panned out dirt from these claims in several places, and finding gold in sufficient quantities to warrant a careful investigation, and knowing that the best way to give it a thorough test was to build sluice-boxes, we moved one of our camps up to this place, which afterward became known as "Robinson's Gulch," and began again the whip-sawing of lumber. In three days we had forty-three feet of sluice-boxes ready for use.

Finding a place where bed rock was visible near the bed of the stream, we located our boxes there. We followed the rock in, and for eight days kept the dirt going through the box. At the expiration of this time we had reached a low spot in the rock, which we thought would be likely to contain a deposit of gold, if it existed here at all in any quantity. But the "clean-up" did not warrant our sluicing any longer at this place, so we abandoned it, and moved our camps back to the lake.

CHAPTER XIII

"MORE COLD FEET"—MINERS GOING HOME

ABOUT this time men began to sell out, and go home in large numbers, and it was no uncommon thing to see from twelve to twenty men, each with a pack upon his back, containing all that he possessed in Alaska, marching along up the trail in single file. At first this brought forth much comment, but it soon became of such frequent occurrence, that when a company was seen going out, all one would hear would be, "More cold feet." The causes which carried so many men out of Alaska were varied.

The first to leave were those who had come never dreaming of what they would be compelled to pass through; and when they came face to face with dangers, hardships, and exposures, they soon weakened, and took the first boat back to the States. Others had gone there evidently expecting to fill sacks full of gold nuggets without much hard work, and in a few weeks be able to return to home and friends, and spend the balance of their days enjoying the fruits of their holiday trip. But when they found that at every step of the way they must come in contact with the hardest kind of work without being able to get any sort of compensation, and only trust to luck for ever getting any, they joined the procession homeward bound.

Others who were anxious to remain would often receive some word from home, telling of sickness, or distress of the loved ones depending upon them, and urging them to come home quickly. All these causes helped to swell the number who were going out. Another element was homesickness. Some had it so bad that no amount of prospects for rapid wealth would have induced them to remain. This condition of things increased until the "home-seekers" exceeded the "gold-seekers."

So many men going home, and throwing so much provision and clothing upon the market, the prices of these articles soon ran down to an extremely low figure. Flour was sold for less than it could be bought for in the States, after all the long and painful task of tugging it in over such a trail. Pork and bacon sold as low as two and three cents per pound; beans brought only seventy-five cents per hundred; and provisions of every kind in about the same proportion.

Clothing suffered the greatest cut, and for two months it would not bring over ten per cent of its original cost. This was a "windfall" for the Indians, for they procured large quantities of clothing, and were often seen parading in the "white man's togs." They also procured from the fleeing miners a good supply of the "white man's eat," and I have little doubt that these red men passed the winter of 1898-99 the warmest dressed and fullest fed of any in all their previous lives.

I should here make mention of some of the berries, vegetation, and flowers which grow in great profusion in Alaska. In spring, as soon as the snow leaves the valleys, vegetation springs up, and grows very rapidly. Indeed so quickly does nature respond to the warm rays of the sun, that it was not uncommon to see full-grown leaves upon bushes which were standing almost surrounded with snow.

In the valleys are many low, marshy places which during the summer months develop a stout, thick growth of marsh hay from three to four feet high. As soon as the snow leaves these places, the grass springs up quickly, and by the first of June a thousand horses could have found good pasturage along the little valley of the Klutina; and during the month of August large quantities of this marsh hay were cut and cured, and put up by these miners into stacks, for winter use. But when haying time came, there were no scythes in this country with which to cut it. Again the old adage proved true, "Necessity is the mother of invention." Several whip-saws were cut up, and converted into scythes; and they answered the purpose, even though they didn't look much like what we Eastern people had been accustomed to see and use. They would no doubt provoke a laugh, if offered as grass scythes at any hardware store in the States, and would no doubt have brought a good price as curiosities, if offered there for sale.

The question is frequently asked, "Can vegetables be raised in

Alaska?" My answer to this would be, that it might be possible to introduce into this climate and soil some varieties which would grow, and under favorable conditions produce a crop; but I have seen the attempt made to raise peas, beans, onions, cabbage, corn, and lettuce, and the result was nearly a failure. I have understood, however, that recent attempts to grow vegetables in other parts of Alaska have been attended with a certain degree of success. But I am fully convinced that any attempt to turn Alaska into an agricultural State must result in failure, unless it should be on the Pacific coast.

Currants evidently thrive in Alaska, for I have seen in the valleys and along the streams which flow through the lowest foothills, large quantities of this fruit growing wild. In taste they resemble those grown in the States, and in size equal those grown under the highest and most favorable cultivation; and we often wondered how much larger they would have grown if thoroughly cultivated. It seemed too bad to see hundreds of bushels of this delicious fruit going to waste, and being of no use to anybody.

Whortleberries, which are so common in the States, are here found growing in the foothills but nowhere did I see them in such profusion as I have in York State, and they seemed to me to be of inferior quality, both in appearance and taste.

But there is a round berry which, except that it is black, closely resembles the whortleberry, and grows upon a very small, low bush, or vine; and this grows in such abundance that I have seen the ground literally black for rods with them. They are exceedingly juicy, and quite sweet to the taste. However, I can not recall an instance where they were relished when first eaten. But we kept on tasting them until in a little time they were eaten with decided relish, and we were soon gathering them by the bucketful for use in the camps. We soon found that they were not altogether a luxury, but were very nutritious; and being so very juicy, they served both as food and drink. Often, while going over the mountains, we would halt, and while resting, gather and eat some of these berries, and find our thirst relieved as quickly and effectually as could have been done at the clearest mountain stream. And often, when out far from camp on the trail, and seeking to find some stream before pitching our tents, our dinner or supper hour would be much delayed; and becoming ravenously hungry, we would stop and eat a cup or two of these berries, and resume our trail without further inconvenience. When stewed by themselves they have a sweet taste not altogether pleasant; but when mixed half and half with currants, they make delicious sauce without the aid of sugar. There are other varieties of berries there, some of which are dry and tasteless, and some juicy and delicious, but not in such great abundance as these.

It has been said that there were no poisonous berries growing in Alaska. How much truth there may be in this statement, I can not tell, but I do know that we ate of almost every kind that we saw growing, and experienced no ill effects from our indulgence.

There is also an almost endless variety of flowers that grow and thrive in this cold climate. Down in the valley and upon the foothills, almost as soon as the snow disappears, the flowers begin to bloom, and from this time until long after the frost appears in the fall, one variety of flower after another follows each other; but the climax is reached in the month of August, when it may be truthfully said that Alaska is a veritable flower garden.

Many times, when out upon our prospecting tours, we would tramp mile after mile through one of Dame Nature's flower beds, and often were they so thickly matted together as to completely hide from view the ground beneath.

I have often stood upon some mountain, and looked away off some ten or twelve miles to foothills which were colored red by the great profusion of flowers which covered them. Away up in the mountains, far above timber, and even above where bushes grow, are certain varieties of flowers and blooming mosses which are in themselves things of beauty. For fragrance, the flowers of Alaska do not compare with those in the States; but for beauty of color, and delicacy of tint, they far surpass those grown in warmer climates.

In winter, the snow falls on the bushes so deep that they are bent and crushed to the ground, and held there so long that they never resume their natural shapes again, but are gnarled and scraggy. Whenever a man went out on a prospecting trip, he was obliged to travel many miles through this brush, the limbs of which were growing in every direction, forming a network so thick in many places as to make it almost an impossibility to get through. Then in pressing his way the small limbs often fly back, switching unmercifully his hands, or face, or eyes; and then often, when so weary as almost to fall under his pack, he might trip his foot in the matted network and go headlong into them. Then, perhaps, to get around some bad spot, he might have to climb for hundreds of feet up some place so steep as to be exceedingly dangerous; then descend into the valley again, only to be compelled to crawl upon a fallen tree across some stream which had become a roaring torrent by the recent rains or the melting snow on the mountains, where if he should miss his hold, he would drop into a stream, the current of which is not less than fifteen miles an hour, to be hurled against rocks which always abound in these mountain streams. And all this under the murderous assaults of countless hungry mosquitoes, which follow one with whetted bills and awful appetites for blood, especially human blood, in swarms outnumbering anything known among keepers of bees; and this for every hour in the twenty-four. No friendly shadows of night put these pests to rest for a respite to their victims, as in the East, or farther South.

This is but one of the features of mining experiences which met us almost daily in our prospecting, and I ask the reader if, in his opinion, it was not enough to try the mettle of almost any man? One can stand up against something large, and fight; but think of being constantly bored by the little insects, and hounded by their incessant "buzz," until it was more to be dreaded than the yell of a panther, or the hissing of a venomous reptile. And is it any wonder that many became disheartened, after several such tramps in unsuccessful hunts for the precious metal, and made a bee line for home?

Boating upon Lake Klutina was a pleasant pastime for us all, whenever we had a few hours of leisure; in fact, this was the only place where we could get away from the everlasting "skeeters" which were almost as omnipresent as one's shadow. Only for these pests, we should have found some pleasure in many of the sights and experiences during the summer months.

Dr. Otaway, of Rochester, N.Y., had brought into this country a very small steam launch, not over fourteen feet long, which had a carrying capacity of from four to six men. And when so many began their journey homeward, the doctor brought his little launch up into the lake, and did a thriving business carrying passengers; and as the shrill whistle of his little steamer resounded from mountain to mountain, it seemed as if we were again back in civilization.

CHAPTER XIV

A GRAND CONCERT - THE INDIAN CHIEF AND

THE PHONOGRAPH

AMONG the camps located on the island were several excellent singers, and it was decided to hold a concert in one of the largest tents; so a committee was appointed and a program arranged.

The evening came, and it was advertised; which, by the way, didn't take long, for whenever a gathering of the miners in any of the large camps was desired, a man would take a cornet, and going to the highest point in the camp, give the "bugle call" three times, and from all parts of the camps the miners would gather to ascertain what was going on. So in this way the concert was advertised, and very soon the tent was filled to its utmost capacity. There were such instruments as the cornet, violin, banjo, guitar, piccolo, accordion, and harp, and we had no lack of men who could handle them in splendid shape.

The name of our own sweet-voiced singer, Mr. Harry E. F. King, of Stamford, Conn., appeared often on the program, and upon this special evening he seemed at his best, and he was encored again and again. He introduced the song, new to us at that time, "On the Banks of the Copper, Far Away," a parody on "On the Banks of the Wabash." The words of the song are here given:

> Round my cold Alaska cabin lies the gold fields;
>> In the distance loom the icebergs, clear and cool.
> Oftentimes my thoughts revert to scenes of childhood,
>> And I wish I were a boy again at school.
> But many things are missing from the picture;
>> Without them it seems quite incomplete.
> I long to store my feet beneath the table,
>> And say once more, 'I've had enough to eat.'

Chorus:-
"Oh, the air is clear and cool along the Copper;
    It's the same in January, June, and May.
Everything is not just what the papers tell you,
    On the banks of the Copper, far away.

"Just one year has passed since I came to Alaska,
    Since I left my darling sweetheart Mary's side;
But to me it seems as if it were a million,
    For from hunger several times I've nearly died.
But I'll try to make my stay of short duration,
    For I long to see my sweetheart, Mary dear;
I also long to see my mother's larder,
    And I'd relish once again a glass of beer."

The entire evening was greatly enjoyed by all present, and will be looked back to in after years as one of the bright spots among so many dark and gloomy hours which made up the greater part of the time in these men's heroic struggle after gold. Much has been said since the war with Spain began about heroic sacrifice, and braving danger, all of which is doubtless true; but after the months passed on this Alaskan campaign, and seeing what I have seen, I can not think that our army and navy have given to the world any higher types of heroes—though they have been by their association rendered more conspicuous—than were developed in the wilds of Alaskan forests, or over Alaskan ice mountains.

A restaurant was opened on the island, and for two months, while so many were on their way back to the States, did a thriving business. It was a great resort for the boys, for the keeper had a phonograph with a large selection of records, both vocal and instrumental.

One day Stickwon,—the chief of the Stick Indians,—with his entire family, consisting of himself and wife and three children (a son twenty-one years old, a daughter of eighteen, and a little boy of five), was on a hunting expedition to the head waters of the Klutina River, and had stopped off a few days at the island to visit the white people. They had killed four caribou on their way up from Copper River, and so were well supplied with fresh meat, a

Chief Stickwon and his family at home. Photo reproduced from first edition.

part of which they traded with the miners for "muck-muck," which is their name for all kinds of provisions.

One afternoon Stickwon and his family called at this restaurant, and being of an inquisitive nature, as all these Indians are, and looking about the place to see what was to be seen, came across the phonograph, and began to examine it curiously.

The proprietor of the place, noticing this, decided to give his distinguished visitor a treat; so calling the entire family, and ranging them around the machine, he put into it a record of a stump speech. Then placing the rubber tubes in their ears, started it going. They all listened attentively, and before it was half through, there was a look of merriment on the face of the old chief. As soon as the speech was ended, he dropped the tubes and fairly danced about the place with delight, saying: "Good! Good! Good!" Then turning to the proprietor, he pointed to the phonograph and said, "How much?"

The proprietor, wishing to have some fun with him, pointed to his daughter, and then at the instrument, and said, "Trade. All same."

A serious expression came over the Indian's face, as he looked first at his daughter, and then at the machine, then at his daughter again. The restaurant keeper began to feel uneasy, for he saw that the Indian was seriously considering his proposed trade, and before the chief had time to make up his mind, the man said, "No sell. No sell."

A look of disappointment seemed to come over his face when he found the phonograph was not for sale.

In speaking of the matter afterward, the proprietor said that he was greatly frightened for fear the chief would take him up at his offer, and then added that he could understand what the phonograph said, but not so with the Indian girl.

Chief Stickwon's son was considered the best hunter and surest shot of any one in the entire Stick tribe. Next day after leaving our camp he demonstrated his title to the claim of being a crack hunter and shot. He was climbing up the mountain in front of our camp in search of mountain sheep, when, upon reaching an elevation of about four thousand feet, he came upon four of these animals a hundred yards or more away. The frightened sheep started up the mountain side, leaping from rock to rock, as only a mountain sheep can, when with a coolness that was remarkable, he raised his trusty Winchester, and gave them four shots in quick succession, and the four sheep fell dead, every one of them shot through the head. This was said to have been only one of the many instances which proved the wonderful skill of this dusky hunter in the use of the rifle.

The little five-year-old son of the chief was very small for his age, and attracted much attention as he went about by himself visiting the various camps, and examining curiously everything belonging to the white man.

One day while at our camp, one of our boys, in a playful mood, jumped at him, and spatted his hands. A frown came over the little fellow's face as he jumped back a few feet, and drawing at the same time a little knife, he assumed the attitude of a duelist, as if defying any one to come near him again. In this position he backed out of the camp, and then ran to their own.

The rifle and knife are the only weapons used by these people, and it became evident that they were taught their use at a very early age.

CHAPTER XV

A RACE FOR CLAIMS

ABOUT the first of August, it became evident by the whisperings about camp, that a "strike" had been made somewhere in the mountains, and that an effort was being made by the discoverers—which, by the way, was no more than natural—to get all their friends favorably located before it became generally known.

Men who were known to have been in the camp at night, would in the early morning hours be missing, and it was no more than reasonable to suppose that they had been given a "tip," and that under the cover of night they had shouldered their packs, and departed for the new strike.

Two weeks before this time a party of our boys had gone out upon a prospecting trip in the same direction in which it was thought the new strike had been made, and a week later another party from our company had left upon a trip in the same direction. I expected to have formed one of this party, but was taken ill the day before they were to leave, and so was left behind.

Two or three days passed without hearing anything definite about the new strike. I had recovered from my illness, and was planning a relief expedition to the first company that had gone out from us (for I knew they must be getting short of provisions), when a courier arrived with a note from one of this party, saying that they had headed off the strike, and believed they were near it, and asked that provisions be sent over the mountains to them.

The next morning I was up early, and taking two men with me we set out on a thirty-five-mile trip, each with a sixty-pound pack. All the forenoon we climbed on up the mountain, through brush and over logs and rocks, until at noon we reached the summit of the narrow divide, where we were to begin the descent into the valley beyond.

Here we came to a beautiful little lake, nestled close to the foot of the mountain, which rose almost perpendicular from its shores to a height of many thousand feet. This lake is the dividing point, and from it the water flows in both directions. We rested here only long enough to eat our lunch, and then began the descent into a valley which we had never seen before.

For the first two miles the divide was so narrow that it might justly have been called a canyon, and for almost its entire length there were fine blocks of stone which year after year had been torn loose by snow-slides and carried down the mountains, until there were enough stones already quarried out and piled along this narrow divide to have walled a city.

Soon we came out of this into the most beautiful valley that I saw in all my travels in Alaska. It was two miles in extent, and covered from mountain to mountain with a network of brush. Through it flowed a clear stream, well stocked with mountain trout.

Looking away to the head of this valley, twelve or fifteen miles distant, several glaciers were in plain view. Looking down the valley ten miles, we saw several hundred acres which were thinly covered with timber and at this point a narrow valley could be seen winding through the mountains in the direction of Tonsina Lake, where, upon one of its tributaries, we expected to find our boys.

We must reach timber before camping for the night, but to do this we were obliged to push on as rapidly as possible, down through the brush which covered the valley. There was an old Indian trail a part of the way, which helped us wonderfully, though in many places the bushes had grown over it so thickly that no trace of it was visible.

About seven o'clock at night we reached first timber, and were so tired that it seemed impossible to go farther. We noticed that a rainstorm was brewing, so we selected a spot as well sheltered from the wind as possible, made a fire, and prepared supper. After supper we sat around the camp-fire and chatted until nine o'clock, when we turned into our sleeping bags, and were soon fast asleep.

How long I slept I do not know, but think it was nearly midnight when I awoke, and the rain was coming down in torrents. In my sleep I had opened the flap of my sleeping bag, and fully a gallon of water had run down inside of it. It wasn't just the nicest place to sleep, partly under water; but it would be no improvement to get up in the drenching rain to empty out, so I arranged my sleeping bag to the best advantage, and lay until morning. There was no need to call me in the morning, for I was out early; and emptying the water out of my bed, I rolled it up ready to march.

The rain continued, and the wood had become so thoroughly soaked that it was only after several unsuccessful attempts that a fire was started, and our breakfast prepared.

After breakfast, though the rain still continued, we shouldered our packs, which were rendered much heavier by being so thoroughly soaked with water, and started across the valley toward the narrow divide in the direction of Tonsina Lake. If there was a dry thread in our clothing when we started, it didn't remain dry long, for the whole valley was covered with rain-soaked brush, which reached to our heads or above, and so thick that with our best efforts we were three hours making the first two miles.

As we reached the opposite side, and entered the narrow valley, we came upon two little lakes, half a mile apart, one containing twenty and the other about fifty acres. These beautiful little bodies of water were upon the top of another divide, and their waters ran in opposite directions.

Passing these lakes, we passed down a small, clear stream, the outlet of the larger one. We were surprised to see thousands of mountain trout flopping about in the little brook, which was only a few inches deep. We shot several of the larger ones, and determined that at our next visit,—if we should make one,—we would bring along our fishing tackle, and see what we could do.

At the lake we came upon an old Indian trail which had evidently been used in former years in going over from Copper River to Tonsina Lake. This we followed for several miles, until it ceased to go in the direction we wished to travel, when we left it again, and took to the brush. Scarcely had we left the trail when we met five men. They had been in this vicinity for several days, and had found the stream upon which the strike was made, staked their claims, and were on their way out to camp.

They told us where the stream was, but gave us very little information concerning it, only that they thought the claims were about all taken. They also told us they had seen the party to whom we were taking supplies, two days before, and that they had crossed over to the head waters of Tonsina Lake.

This was a surprise to us, that they should have gone so far away. We afterward learned that they had been informed about the strike, but their informant had either made a mistake, or they had misunderstood him, and they received the impression that it was upon a stream emptying into the head of the lake, instead of the foot, and so had gone over there.

The little stream down which we were going emptied into the one upon which the strike had been made, but we were several miles from it. We knew that the last party of our boys, which had left camp the week before, was somewhere in this part of the country, and we pushed on hoping that in the early afternoon we might find them; but if not, that we might yet be in time to get claims.

The rain, which had been steadily falling all the morning, had ceased, and the sun came out; and the mosquitoes came too. They had given us a rest during the storm, for they can not get in their work during a hard rain; but they were here in full force now, and, seemingly intent on making amends for lost time, they succeeded in making it very interesting for us the remainder of the day.

Speaking of mosquitoes, it was reported, on what seemed good authority, that upon one of these prospecting trips, some miners had come upon the dead and swollen bodies of four men. It was believed, from what could be ascertained, that they had lost their helmets by having them torn in passing through the brush, and having no material to replace them, had fallen victims to the rapacity of these hordes of merciless blood-suckers, and been actually poisoned to death.

The noon hour had arrived when we reached the top of a steep bank, down which we must descend a thousand feet or more to reach the bed of the stream. We had sat down for a few moments of rest, when we heard the cracking of brush below us, and looking down, we saw what we supposed to be two men climbing up through the thick brush toward us; but on their coming up to us, we found that one was a woman dressed in male attire. They were a man and his wife who were camped near us on the island,

and had been down on the creek staking claims. They informed us that only a few hours before they had parted with five men belonging to our company, and that they had also staked claims, and were then camped three miles down the creek. This was welcome news to us, but they told us also that the trail leading down to where they were camped was something terrible.

We had now heard definitely from both our parties, and expected to have no difficulty in finding them. Clinging to bush and rock, we climbed down to the bed of the creek, where we saw a stake upon which was written, "Discovery Claim." The stream was too deep to ford, and the water was rushing toward the lake at the rate of twelve miles an hour, dashing itself into foam upon the rocks which everywhere lined the stream.

We only halted at "Discovery Claim" long enough to make and drink some coffee, when we shouldered our packs, and resumed our march down the stream. We had not proceeded far when we reached a point where the current of the stream came in close to the foot of a perpendicular wall of rock fully five hundred feet high, and the only way we could get below was to retrace our steps for some distance, climb up the steep mountain side to the top, and make the descent again to the valley below it.

I shall not soon forget my experience in passing this spot. I had reached a point perhaps four hundred feet up from the creek bed, when I thought I could climb around the rocks to a place where I could begin the descent; this, if it could be done, would save me a climb of several hundred feet both up and down, and I had reached the most difficult part of it, when the rock, which was of a shaley character, began to crumble from under my feet. I clung to the rocks above me as best I could, but their crumbly nature made them very unreliable. I gave one glance downward, and there, hundreds of feet below were the rushing waters, into which I must certainly fall if I once got started. I thought to loosen my pack, and let it drop into the stream below, to lighten the burden on the shaley rocks, and relieve me of its unwieldy proportions in such a tight place as this; but I could not let go my hold to unbuckle the straps. To go further I could not, and it seemed almost as difficult to get back; but I plainly saw that was my only chance, so I began working my feet along a few inches at a time, expecting every moment that the shalely rocks under me would give way, and precipitate me into the mad waters below. Little by little I retreated from my perilous position, and reached a place of safety, so weak from the great strain upon my nerves that I could scarcely stand up under my pack.

After a brief rest I followed my companions, who had climbed higher up to a place where the mountain was not so steep, and we reached the valley again in safety. To pass such places as this with heavy packs taxed one's strength to its utmost, and keeps the nerves under a fearful tension.

I'll not attempt to describe in detail the balance our trip over this three miles, but four different times we were obliged to climb the high mountain to get around some rock which projected out into the stream, and it was four o'clock when we came in sight of our boys' camp.

They had been out sinking some holes to test the claims which they had staked out the day before, and had but recently returned to camp, and were just preparing their dinner. One who has never been separated from home and friends, and in such an isolated country as this, can never tell our feelings as we emerged from the bushes, and saw only a few rods distant from us this, to us, homelike scene. We were ravenously hungry, and there before us were skillets of frying bacon, and stacks of steaming "flapjacks," and buckets of hot coffee. We were so fatigued that we could scarcely drag one foot after the other, and here was a roaring camp fire, where we could dry our soaked sleeping bags and clothing, and plenty of spruce boughs to make us a good soft bed on which to sleep.

When the boys saw us come out of the brush, they left their seats by the camp fire, and rushing up, shook us by the hands so heartily that a looker-on might have supposed that we were brothers who had been separated, for long years, instead of only one short week. We sat around the fire, and recounted to each other the various experiences of the past week, until late in the evening.

They were camped two miles from Tonsina Lake, upon a creek which emptied into the lake one-half mile from its foot, and which had been staked along its whole length to the marshy approaches to the lake. I determined to visit the head waters of this creek and if possible get claims there; but before doing so we must find our other party, which we knew by this time must be nearly out of provision. We expected they were near the head waters of the lake somewhere, and might possibly be camped on its shore. So in the morning we went down where we could get a view up the lake for fifteen miles, believing that if they were upon its shore, we could see them, as they had a small tent with them. And about ten miles up the lake we could plainly see a small tent, close to the mouth of a gulch. We felt sure that this camp belonged to our boys, and tried to attract their attention. We made a large fire on the beach, and also set on fire the tops of several large spruce trees, which flamed up a hundred feet or more into the air.

We waited some time, but they gave us no sign that they had seen our signal fires, and we thought the surest way would be to send some men to their camp. This we knew would be a hard trip, for the shore of the lake was marshy, and covered with a dense growth of brush, through which those who went must push their way, accompanied by clouds of mosquitoes, the like of which we had not seen before.

Not wishing to select any one myself to make the trip, I called for volunteers, and two men agreed to go. It is worthy of mention that during all the seven months in which I was superintendent of the company there was never a task to perform which was accompanied with extreme danger, or a trip to make which must bring to those who made it great fatigue and exposure, that there were not more men volunteered to do it than were required for the work.

On the shores of the lake, near where we had made our signal fires were the remains of several log buildings. They had been built long years before, but their surroundings gave evidence of having been occupied not many years ago. The larger portion of them had been burned, and those which had escaped the fire had been so completely torn down that scarcely one log was left upon another.

In one corner of what had been the larger building were several tons of quartz, which was broken up fine, and had evidently been exposed to extreme heat. We came to the conclusion that whoever had occupied these buildings had found a quartz ledge containing gold somewhere in the mountains near by, and that they had taken it to this building and put it through a miniature smelter to extract the gold.

Who it was that had occupied them we never knew, but scattered around the yard, which was covered with grass four feet high, were the remains of several varieties of birch bark baskets, and also hats made from the same material. This was evidently the work of Indians, and many of the stumps in the cleared space about the buildings, from which the trees had been taken to build them, gave evidence of having been cut by Indians, while others were the work of white men.

It might be well to tell the reader how we could tell whether a tree had been cut by a white man or by an Indian. It was in this way: a white man cuts a tree on two sides only, while an Indian cuts it all around. We thought the work had been mostly done by Indians under the direction of white men, but why the buildings had been so completely destroyed was always a mystery to us.

After our two men had departed on their trip down the lake, and we had thoroughly inspected the site of the old buildings before mentioned we returned to our camp to await their return, which we expected would be about noon next day. It was past two o'clock when we saw them coming. We had prepared for them a sumptuous dinner, and they were in condition to enjoy it, for they had been on short rations for two days. The remainder of the day and evening was spent in relating the adventures of the past two weeks, and in making plans for the future.

It was decided that those who had already staked their claims should return to the island, leaving all their provisions with those who stayed, except enough to last them back to camp. Those who had not staked were to go to the head waters of the creek, and if possible get claims.

So after an early breakfast, and bidding our five men who were to return to the lake good-by, we set out up the stream. The trail upon the opposite side of the stream was said to be the best, but to get across the swift current was no easy task. We were determined to make the attempt, however, so going some distance up, we found a place where a tree could be felled to reach across. It took us some time to chop it down with our hatchets, but we soon had the satisfaction of seeing the stream spanned. The stream at this point was not more than thirty feet across, but was very deep, and ran at a tremendous rate.

If any one thinks that to cross a stream like this on a springing sapling, with a seventy-five-pound pack on his back, is not a difficult thing to do, he has something yet to learn. It can be done, and we did it; but the first man over came near falling off into the water, and reached the opposite bank so frightened that it took him some time to recover his nerve. After this we cut a long pole, and while a man was crossing, two others stood a little way below holding it down to the water, so that in case he should fall he would stand some show of being rescued. We had all succeeded in crossing safely but one man, who was not a member of our company, but was accompanying us, and he said that no prospect for gold, be it ever so good, could induce him to place his life in such peril. So he left us, and returned to his camp at the island.

We followed up the stream until we reached "Discovery Claim." Above this, for two miles, the stream flowed through a narrow rocky canyon, the walls of which were thousands of feet high, and to get up the stream from where we were, it was necessary to climb over the top of these mountains, and come down again into the valley above the canyon. It looked like an almost impossible task weighted down as we were with heavy packs; but there was no other way,—so we began the ascent. Often we were obliged to sit down and rest; then would go on up an incline so steep that to look back would almost make one dizzy. After nearly two hours of such climbing we reached the top, and took time to rest. From this lofty point we had a fine view of the surrounding country. Down the valley of the Tonsina we could look away beyond the Copper River, fully eighty miles distant, to where could be plainly seen the snow-capped peaks of Mounts Tillman and Blackburn.

But we had little time to devote to scenery now so we continued our journey, following along the crest of the mountain until we came to a point where we could make the descent, which we did without mishap. Upon reaching the stream again, we saw that it had been staked; so we continued on up for three miles, where night overtook us, and we struck camp.

CHAPTER XVI

EXPERIENCE WITH ALASKAN BEARS

WHILE following the creek, we saw in the sand some of the largest bear tracks we had ever seen. It seemed to us almost impossible that there could be bears in this country large enough to have made these enormous tracks. But here they were in plain sight. We measured some of them, which were eight inches in width and sixteen inches long. Some of the boys said that if bears grew to such a size as these tracks indicated, they hadn't lost any, and were not looking for any.

We had gone into camp above where the creek had been staked, but just how far we did not know, for it had become dark. So we made a large camp fire, prepared supper, and went early to bed. In the morning we were up early, and leaving one man to get breakfast, started out to find the last claim which had been staked, and then to put down our own stakes. During this trip this incident occurred:—

Two days before, when we reached camp near the foot of Tonsena Lake, there was in company with us a young Irishman, whom I will call Tommy. He was a persistent prospector, and had been with us on more than one long tramp, but as yet he had never had an occasion to put down any stakes. He had talked much about the new strike, and seemed anxious to get there and stake him out a claim. But we had reached the camp too late for Tommy to get his claim staked. However, after supper he went out and cut two stakes weighing ten to fifteen pounds each, brought them to the camp, squared the tops, and wrote on them that he was a citizen of the United States, was over twenty-one years of age, and that he had this day located the placer mining claim described below. Then he gave a description of what his claim would be, and signed his name.

"Now," says Tommy, as he viewed these marked stakes with much apparent pleasure, "Oi'll hov' thim riddy for mornin'." In the morning Tommy was the first man ready, and stood around with his stakes upon his back for some time.

Soon we started for the lake, where we intended to stake claims, if we found any desirable ones left. All the forenoon, while pushing our way through brush and swamp, Tommy carried his two stakes upon his back, and, to our surprise, brought them back to camp with him at night. The next morning we knew that there was before us a hard tramp of from eight to ten miles, with a high mountain to go over, and imagine our surprise to see Tommy appear with his two stakes again over his shoulder. I asked him what he expected to do with these.

"Oh," said he, "Oi think Oi'll take me two sticks."

"Why, Tommy," I said, "don't you know that we are to have a long, hard tramp, and you can find stakes anywhere, as well as to carry these along?"

"Yis, but Oi hev these all riddy," was his reply, and no amount of argument could persuade him to leave them behind.

He carried them all day; and, after we had camped, and a fire had been started, Tommy picked up his stakes, saying, "Oi think Oi'll go and stick me stakes." Whereupon he started out into the gathering shadows of evening down the stream, but he had not been gone long when I imagine he remembered the monstrous bear tracks we had seen during the day, for he soon made his appearance again, and throwing the stakes on the ground, remarked, "Be gabs! Oi'll wait till mornin'."

The next' morning Tommy shouldered his stakes, and accompanied us down to where the last claim had been staked, and on our arrival said, "Now, shill Oi stick my stakes first ?" I informed him that he might if he desired, so the stakes were set, and Tommy had rest, after carrying fully twenty pounds of wood for at least fifteen miles.

That morning we staked nine claims of twenty acres each, thus making fourteen in all held by members of our company on this one creek. This we considered to be enough, if they should prove any good, and surely enough if they were worthless. The balance

"We found good surface indications, in fact the best colors we had struck in all our prospecting; so we determined to return to the creek, put in a sluice box, and give it a thorough test." Photo reproduced from the first edition.

of the day was spent in panning out dirt taken from the creek at different points on our claims. We found good surface indications, in fact the best colors we had struck in all our prospecting; so we determined to return to the creek, put in a sluice box, and give it a thorough test.

The next morning we made preparations for returning to camp. Leaving all our tools, and all the provisions not needed on the road, in a cache under a tree, for use when we came back, we set out for Lake Klutina. Just as we were starting, we met a large company coming in to get claims, and gave them all the information we could. We afterward learned that they staked twenty-five claims on the creek above ours. Our return trip was uneventful, and on the morning of the second day we reached camp on the little island. A few of our men who had been left there had been getting discouraged with the outlook in that part of the country, and about this time five of them decided to return to the States. They were N.D. Benedict, Bernard Gasteldi, W.H. Lawrentz, Fred Gittner, and John Potts.

Our association with these gentlemen had been of the most agreeable character, and their decision to return to their homes was much regretted by the entire company. Mr. Potts had for some time been in charge of our store at Valdez, and his manner of conducting the business at that point had won for him the admiration of all. A satisfactory settlement was effected with them, but as it would be some time before another boat would leave Valdez, they were to make their homes with us for several weeks, and during that time make a trip down to Copper River.

The company was now reduced to sixteen men, and as the time had arrived for our semiannual election of officers, and the company being now all together, it was decided to hold the election at this time. I was elected vice-president, and re-elected as one of the directors, and also to a second term as superintendent.

About the middle of August two hunters returned to camp one night, and reported that they had wounded an enormous bear only a few miles from camp, but had become so frightened at the actions of the infuriated animal that they had returned to camp with no intentions of going after him again.

An old bear hunter from Texas happened to be there, and he decided to return next morning, and look for the wounded beast. So in the morning he took his rifle, and with a small sailboat started across the lake to a point two miles above, where our hunters of the previous day had told him to leave his boat. They had also given him minute directions as to the course they thought it best to take to find the wounded bear.

He had proceeded only a mile, after leaving the boat, when suddenly, and without warning, the bear stepped from behind a tree, and, as the hunter was in the act of raising his gun, struck it with his huge paw, and sent it flying through the brush many feet away. Then he proceeded to handle the man very much as a cat would handle a mouse, cuffing and rolling him about on the ground. With his teeth he tore the flesh completely off both his cheeks, and crushed both his jawbones in a horrible manner. Then the bear, seeming to be satisfied with his revenge, left him.

The hunter, more dead than alive, managed to crawl to his boat, clambered into it, headed it toward camp, and fainted, falling across the tiller, thus holding it on its course. Some one saw him coming, and went to meet him, and was greeted with a sight that almost curdled the blood in his veins.

Medical aid was immediately procured, but it was little that could be done for the poor fellow. However, he recovered consciousness, and not being able to talk, wrote out how it had happened, as before stated. He said that the bear was of enormous size. To all appearances it had heard him approaching, and, standing upon its hind feet behind a tree, had waited his coming, letting him approach so near as not to give him a chance to use his gun; which was evidently the case, as the weapon had not been discharged when it was found. The unfortunate man was made as comfortable as circumstances would permit, but he lingered six days, and died.

The same thing came near being repeated two days afterward. Two men who were out hunting came across a large bear, and wounded it; whereupon the enraged animal charged the men, knocked one of them down, and was pawing him around on the ground as the other bear had done, when the other man rushed up, and placing the muzzle of his gun to the bear's ear, fired. A great quiver passed through his immense frame, and he fell dead upon his intended victim.

The men who had this experience were strangers to me, but one of them was pointed out to me a few days later as being the one who shot the bear, and I said to him: "They tell me that you are the man who shot the bear the other day."

"No, sir," he replied, "I am the man who held him while the other man shot him."

These are but two of the many blood-curdling adventures which the miners had with these terrible brutes, which are said to be, when wounded, a worse enemy than the famous grizzlies. I heard of one instance where eight men, while hunting, came upon one of the largest of this kind. The bear charged the men right and left, with no less than twelve bullets lodged in his body, but was finally killed by a charge of buckshot fired at a distance of only a few feet from his head. His skin measured eleven feet in length, and the entire eight men slept on it that night.

We remained at camp a few days to rest, when a party was made up to return to the scene of the new strike, with a whip-saw outfit and tools for building and operating a sluice box. There was so much to be carried that we could only take a limited amount of provisions, so it was planned that five or six men should be, kept at work carrying provisions to the working party. Our packs were made up of sixty pounds each, and we again set out over the mountains.

Before noon of the second day we reached the trout stream which we had passed on our previous trip, in which we had seen such an abundance of fish. Near the head waters of this creek we found a place where some one had camped but a little while before, and there in one pile were nearly half a bushel of trout heads. So we knew that some one who had preceded us, had made a good catch.

We determined to try our luck also, and shooting a bird to use for bait, we began fishing, and in a short time had caught upward of forty pounds of the speckled beauties. These were taken along with us, and served as a valuable addition to our food supply.

The next morning five men returned to the lake after more provisions, while the others began the work of whip-sawing for our sluice boxes.

Every day large companies of men passed us on their way up the creek, hoping yet to be able to get claims. But the creek was staked along its entire length by this time, though it was twenty-five miles long. This creek upon which we were working also contained large quantities of salmon as well as brook trout. The salmon were caught with a hook made with a spike fastened to the end of a pole, as previously described.

CHAPTER XVII

FLOPPED BY A BIG SALMON

ONE day, while standing on the bank looking for these fish, I saw a large king salmon swimming leisurely along up the stream against the strong current, but too far out to be reached. I walked along for some distance, keeping opposite the fish, when he suddenly came in toward the shore. I stepped nearer the water's edge, when he again turned toward the center of the stream. Reaching out as far as I could, I hooked him near the tail. The result was a surprise for me, for he made a sudden lunge away from the shore, and I was jerked instantly from under my hat, which fell in the water, and but for a large rock I would have landed in the creek also; but bracing my foot against the rock, I succeeded in landing the fish, which only lacked one inch of measuring four feet in length.

At the end of five days we had fifty-six feet of sluice box ready to be put in place. The same night, the party who went back to the lake after provisions returned with a good supply, and also brought in our mail, which had arrived after we had left, and which we were always so anxious to receive.

For some time we had desired to go on a prospecting trip to the head waters of Tonsina River above the lake. It was next to impossible to reach this point by going up the river, for it emptied into the lake through a rocky canyon several miles in length. A few people had made the attempt, but had given it up as a hard and dangerous undertaking. We had planned to prospect a few of the gulches coming from the direction of the Tonsina, and emptying into the head waters of the stream upon which we already had our claims, partly in the hope that we might find a quartz ledge in one of them, and partly that we might find a passage through to the Tonsina valley.

Next morning the sluice box was put in position, competent

men were put in charge of it, and a few men started back to camp for more provisions. Then H. H. Sweet, Charles Priceler, and I started out upon this trip, taking with us only a few days provisions.

The first day we reached the head waters of the creek on which we were working, fifteen miles from camp. We passed the night at the mouth of a deep gorge, which we decided to explore next morning. To enter this gorge, we must climb a thousand feet or more, up a steep mountain, at which point it seemed less than two miles to the head of the gulch. So leaving our packs at the camp, and taking with us only our dinner and prospecting outfit, we started out, intending to return at night. After traveling an hour we became convinced that instead of being two miles it might be eight to the head of this narrow valley. We passed several small lakes, the waters of which were as blue as the sky overhead.

Upon reaching the upper end of the gorge we discovered that it was only a basin, with no outlet to the valley beyond. There was a glacier two miles in extent at the upper end of this basin, and in the rocks at the farther end of this we saw what appeared to be a quartz ledge. We climbed anxiously this two miles up over the ice to this ledge, only to find that instead of being quartz, it was simply a vein of white porphyry rock.

Not having found upon this little stream any indications of gold, or any passage to the valley beyond, we returned late at night to the place where we had left our packs and were to stay for another night. The work of the day ended in disappointment, but was only one of many similar days which had ended similarly, not only to us, but to all prospectors after the yellow metal, especially in Alaska.

Two miles farther on was another narrow divide leading off toward the Tonsina, and we hoped to discover a passage through it, to the head waters of this stream.

The next morning we shouldered our packs, and keeping along the side of the mountain for two miles, we entered the divide. Mile after mile we traveled, stopping occasionally to examine some narrow vein of quartz, or pan out some dirt to ascertain if it contained gold, until one o'clock, when, to our joy, we found ourselves standing upon the summit of a narrow divide which led through into the valley beyond, down which we felt sure must flow the Tonsina.

However, to make sure of this, we left our packs at this point, which must have been at least three thousand feet higher than the valley which we had left in the morning. Going down the divide, we passed over snow which was from twenty to forty feet deep, and upon which were tracks of bear, moose, and other animals made while passing from one valley to another. Three miles farther on we came out upon the top of the foot-hills, two miles distant from Tonsina River, and fully three thousand feet above it. Looking down the valley ten miles we could see the upper half of the lake, and looking above us fifteen miles several glaciers at the head waters of the river were visible.

We decided at once to go back to camp, get ten days' supply of provisions, and return again to this stream. So making a forced march, just before dark we reached the spot where we had stayed the two previous nights, and the next morning early were on our way to camp, reaching there in the middle of the afternoon.

As yet there had been no mining district found there, and the creek upon which our claims had been located had never been named. But a meeting of the miners holding claims upon this creek had been called, and was to be held upon "Discovery Claim" the next day. So at the hour appointed, eight of our company were present. Upon arriving there, we found about forty-five men gathered in front of a brush house which had been used as a camping place by the company making the discovery. At this meeting the stream was named "Manker Creek," in honor of the man making the discovery; and he was also elected recorder of the district, which embraced all the drainage of the Tonsina valley.

After the meeting we returned to camp, and the next morning Sweet, Priceler, and I, taking provisions for ten or twelve days, set out on our return trip to the head waters of the Tonsina. The first day we reached the last timber on Manker Creek, where we camped for the night, and had scarcely gone into camp when it began to rain.

As soon as supper was over, we crawled into our sleeping bags; but it was little sleep we got that night, for the whole night through it rained hard, and the oilcloth cover forming the outside

Margeson's claim "No. 19 above the canyon on Manker Creek".
Tonsina Mining District Records, p. 46.

had become somewhat worn, so that our sleeping bags were
soon soaked with water.

I hope no one will envy us our pleasurable feelings as we
slung our water-soaked packs upon our wet shoulders next morning,—packs which were heavy enough before, but were now almost twice as heavy for being wet,—and started up the divide
against the storm, which was driven in our faces by a gale so
strong at times that we could make no headway against it.

Having fifteen miles to go before reaching any timber for a
camp fire, it was necessary for us to push on as fast as possible. We
came upon several flocks of ptarmigans, which were so tame that
they only ran out of our way a few feet, where they would sit and
quack like young ducks. We shot several, and tied them to our already too heavy packs, intending when camp was reached to
make one of those delicious ptarmigan stews, which the boys declared was a dish fit for a king.

About noon we arrived to within about two miles of the summit of the divide, and over this distance was perhaps an elevation
of eight hundred feet to the mile.

CHAPTER XVIII

EXPERIENCE WITH "WOOLEYS"

OUR route lay over large rocks which had tumbled down from the mountains, and we were compelled to travel in one of the fiercest storms of wind and rain that I had ever experienced.

The "wooleys," as they are called in Alaska,—which, by the way, travel something like a whirlwind,—go sweeping along the mountain side with such force that great rocks are torn loose by them, and sent tumbling into the valley below.

One peculiarity about the "wooleys" is that they will come rushing along down the valley at a terrific rate, then, without any apparent cause, turn at right angles, and go straight up the mountain for thousands of feet; then again turn as suddenly, and go tearing along the side, or go headlong down into the valley below. And they are of such frequent occurrence during some of these storms that a person can not stand in any one place a minute without being struck by one, and their liability to come from any direction made them more difficult to cope with.

As we pushed on up toward the summit, these "wooleys" became more frequent and of greater force. We were not only in danger from them directly, but took chances of being crushed by rocks which were loosened and sent tumbling down the side of the mountain as they swept this way and that. By keeping close watch we were enabled to note their approach in time to throw ourselves flat on the ground, where we would cling to the rocks until they had passed, and then rise and move on as fast as possible; but we never went more than fifty feet before having to seek the ground again.

Getting up and down so often with a pack of eighty pounds or more on one's back was a task which the strongest could not endure long, and by the time we had passed over the summit we were so thoroughly exhausted as to be obliged to stop for rest. Finding a place between two ledges that was somewhat sheltered from the "wooleys," we sat down to rest, and watched them as they went rushing back and forth across the face of the mountain.

Exactly in front of us, and perhaps two thousand feet above, was a little stream, which, swollen by the recent heavy rain, was pouring over the rocks, making in its last leap a waterfall of about one hundred feet high. One of these "wooleys" came sweeping along the mountain side until it came to this stream, and then suddenly turned and went straight up toward the waterfall. We were all attention, to note what it would do then, and were surprised to see it pick up the water and carry it back up the mountain for hundreds of feet above the falls, and so fierce was the wind that for nearly a minute not a drop of water came over the falls.

After a brief rest we resumed our journey down the divide. When within two miles of the Tonsina, we wished to cross the foothills and strike the river farther up, but the rains had turned the little mountain stream into a roaring torrent impossible to cross.

We could not follow the stream down, because of its narrow rocky walls, so going off to one side we entered one of the most impenetrable jungles I ever saw. The brush was ten feet high, and so thick that but for the fact that we were going down hill we never could have got through.

This was kept up for some two miles, when we came out into a dense forest of spruce timber. As night was approaching, we began to look about for a camping place. Following up through the timber we soon came to the stream which we had been unable to cross up near the top of the foothills. Here in the heavy timber we made a large camp fire, but it was impossible to dry our clothing, as the rain still kept falling, and we knew that with wet clothes, and our sleeping bags soaked with water, a very uncomfortable night was before us, and so it proved.

We were more than glad when the first streaks of dawn appeared, and the rain having ceased, we hung our sleeping bags up around a roaring camp fire to dry, and also stood around it to dry our soaked clothes. We had intended to go on up the river, but it had risen so from the recent rain that it was bringing down logs, and even whole trees; and standing upon the bank, we could hear large boulders being tumbled along on its bottom by the rushing current. So we decided to spend this day in prospecting and hunting.

CHAPTER XIX

SHOOTING A BEAR

WE had with us one Winchester rifle of large caliber, and one Winchester shotgun, but unfortunately all the shells for the latter were loaded with No. 8 shot, which were all right for ptarmigans and spruce hens, but if we should chance upon a bear, they would be of little service. Taking our guns and prospecting outfits, we followed down to the main river, which I believe in low water could have been forded, but now would have floated two loaded canal boats side by side.

On the bank of this river was a hard beaten path, made by wild beasts as they traveled up and down the stream. There were large numbers of bear tracks, and now and then a half-eaten fish would be found where some of these forest monsters had made a meal, and seemed to have had a surplus. Soon we came to one which appeared to have been recently caught, and only a few mouthfuls had been eaten from it. The boys declared that we were surrounded by bears, and one of them said that he wouldn't mind hunting them, if he knew we wouldn't find any.

We followed along down the river half a mile, when we came to a rocky ledge which prevented us from going farther. Noticing a cleared place to our left, we went to examine it, and found it a marshy spot containing ten acres or more, through which flowed a small, clear stream of water.

I was in advance, and looking across to an elevation on the opposite side about two hundred yards distant, saw a large brown bear emerge from the thicket, and stop in full view of us. His nose was upon the ground, as if eating berries, and to all appearance he was unconscious of our presence. My companions saw him at about the same time that I did. I was carrying the Winchester, and quickly raising it, took careful aim at his heart, and fired.

For an instant his bearship performed several acts which would have done credit to a circus-trained animal, and then disappeared behind the mound. We crossed over to where he had stood when I fired, and following his track a short distance, saw where he had entered some brush. The bushes were covered with dark red blood upon the side opposite to the one I had shot at, so it was evident that the bullet had passed through his body.

Should we follow him into the thick bushes? was the question which now presented itself. All the blood-curdling stories we had heard about these animals when wounded came up before us, and we knew well enough that at such times they are terrible enemies to encounter.

We had practically but one gun, for the smallness of the shot in the shotgun rendered it of little service in a battle royal such as this promised to be, provided we should find him still able to fight. However, we needed that bear, and our anxiety to get him overcoming all our fears, we decided to go after him, and take our chances. So taking the lead with the rifle, and my companions a close second,—one with the shotgun, and the other with a miner's pick, shovel, and gold pan,—we followed the bear into the dense jungle.

The bushes were so covered with blood that it was easy to follow his trail, but so thick were they that at no point could we see more than twenty feet ahead of us. Carefully we parted the bushes, and examined every particle of ground, taking only a few steps in advance at a time, for we did not propose to be caught napping, or be taken by surprise, even if we were pursuing such a wily foe. Our knowledge of their habits, from our recent acquaintance with what they had done to some of our fellow miners, was too fresh in mind to let us forget.

We soon found where he had lain down, and it appeared to us, from the indications of his trail, that he could not proceed much farther. After following him a hundred yards or so, we came to a small hollow in the ground which contained a few barrels of water. Into this he had gone and washed himself, coloring the water red with his blood, but in some manner had stopped its flow. However, as long as he remained in the thick bushes we had no difficulty in tracking him, but he soon came out upon an old creek bed, where it was impossible to tell in which direction he had gone.

We spent an hour hunting for him, but without success. How we wished for a little snow just then, that we might tell in which direction he had gone, or for a dog to follow his trail, for we were convinced that he had received a death shot, and was not far away. We were therefore compelled, very reluctantly it is true, to abandon the chase. But our guess proved correct about his having received his death shot; for some hunters, a few days afterward, came across his dead carcass only little distance from where we had lost his trail. He was indeed a monster, for they said that he would have weighed at least eight hundred pounds.

Of course we were grievously disappointed at losing him, when he seemed almost sure to us for it was not the loss of the meat only for which we cared, but to bag such large game would have been something worth repeating to our children and grandchildren in future years. Yet we were not altogether forgetful that our misfortune may have been our safety, for even wounded as he was, he might have made it exceedingly lively and interesting for us. Others of his kind had fought with awful desperation with many bullets in them, and he had but one. The very severity of his wound might have added to his desperation, and even given him unnatural strength for a last struggle.

In recalling this adventure to mind, after having learned more of the ferocious nature of these animals, I can but believe that while pursuing this wounded bear through the thick jungle, our lives were in greater danger than if we had been shooting the great Klutina River rapids, where it was said that three out of every four boats that made the attempt were wrecked; for had we met him, and failed to kill him instantly with the first shot, the chances no doubt would have been greatly in favor of the bear.

While crossing the marsh where the bear had been shot, we saw in the small, clear stream, swimming leisurely about, a large number of salmon; and now that our bear hunt was declared off, we decided to go fishing. So returning to our camp we procured our salmon hook, or spike, which we always carried with us on our prospecting trips, fastened it to a pole, and returned to the stream, and within fifteen minutes had caught twelve salmon whose average weight was not less than eight pounds. These were strung upon a pole and carried to our camp, a very considerable load for two men, and fresh fish formed the staple of our menu for the time we remained at this place.

We prospected in different places as deep as the high state of the water would allow, and washed pan after pan of dirt, but with very little encouragement.

During the day our sleeping bags had become dry, and we passed a comfortable night in them, the first since the great storm began. When morning came, the water in the creek had so abated that we were able to cross and continue our journey up the Tonsina. The water in the Tonsina, however, had receded but very little, and in many places was still over its banks, which compelled us to keep along the foothills, over trails which were too bad to describe. Sometimes we were obliged to spend hours in bridging streams which were too deep to ford, and over which there was no other possible way of crossing.

Day after day we pushed on up the valley, stopping here and there to prospect the small side streams which empty into the main river. At the end of the sixth day out we came within a short distance of the glaciers at the head of the river, and stood in one of the wildest regions it had ever been my fortune to enter. Here we built us a small brush house, in which we slept three nights, and during the days prospected the surrounding country. We found some gold, but not in sufficient quantities to warrant us in staking claims, and as our provisions were beginning to run low, we were obliged to commence the homeward march.

To get out of the valley of the Tonsina was harder than to get in, and we had found that no picnic. By reaching the foot of the mountain, we might work our way along above the brush line to the divide by which we had entered the valley; but to the upper edge of the brush from where we were was four miles, at an elevation of 3,500 feet, and every step of the distance must be made by pushing up through a perfect network of bushes as high as our heads. The coming down had been an undertaking to be dreaded, but it was nothing to be compared with making the ascent.

Getting an early start, we began our return trip next morning. We had more than twenty miles to travel to reach the nearest timber across the divide, and whether we would be able to reach it before night or not was a question we were not sure about.

I shall never forget that four miles of travel. Many times we were on the point of giving up and saying we could go no farther, but after a brief rest we pushed on again, and at last reached the foot of the mountain. Here we sat down, and took a long rest; and the reader may be assured that, with us, the desire to see the head waters of the Tonsina had been gratified forever. We had seen it. Its wild jungles had been prospected, and we were more than satisfied.

Nothing of note occurred during the day, and at seven o'clock we reached first timber on Manker Creek, so tired that some of the boys said that if it had been another half mile they could not have made it.

We had camped several nights at this same spot, and had christened it "Ptarmigan Camp," on account of the large numbers of these beautiful birds which always frequented this particular spot. I have given a brief description of these birds in a previous chapter. Just before reaching camp we had shot several of them, and after supper we dressed them, and hung them over the fire to prepare them for breakfast next morning.

Being very tired we turned in early, and had a good night's rest. I was awakened early next morning, almost before daylight, by what seemed the quacking of hundreds of ducks, and rising up in my sleeping bag, I saw above us, around us, and on every side of us an immense number of these ptarmigans, which were flying about and over us entirely unconscious of our presence. Coming in so late at night they had not been apprised of our arrival, and our little sleeping bags, spread upon the ground here and there, had not attracted their notice.

I remembered that we had but four shotgun cartridges left, and did not forget that it was forty miles to where we could obtain more; notwithstanding which I could not restrain my desire to have some fun with this gabbling flock, and crawling out of my bag as quietly as possible, I caught the shotgun, and blazed away among them, bagging four.

You should have seen the boys, when the report of the shots woke them from a sound morning nap. How they glared! They rose quickly up in bed, and looked about them wildly, as if they thought we had been surprised and fired upon by some wild Indians. The birds, to whom I suppose the firing of guns was unknown before our advent among them, seemed not to realize any danger, and scarcely noticed my sudden onslaught upon them, but remained in our vicinity for some time.

Breakfast over, we set out for our next camp, fifteen miles farther down the stream. During our absence this stream had also overflowed its banks, and on our arrival we found that the boys had not been able to sluice much on account of high water, and leaving their tools and a small amount of provisions in the tent, had returned to Lake Klutina.

CHAPTER XX

ANOTHER MAD RUSH FOR CLAIMS

WE remained in the tent overnight, and in the morning started for the lake also. We had made about half the distance—say fifteen miles—when we came upon a camp fire, around which were several men preparing themselves coffee; and while resting there they informed us that a new strike had been made somewhere near there, and that they were on their way thither. They also informed us that a large company were coming behind them.

Wishing them luck we started forward, but the sight and smell of steaming coffee had made us hungry, and so we decided that we, too, would have our dinner. We had barely got it prepared, and begun eating when three men went past us down the trail at a two-forty gait. We asked them what was up, but they did not seem to have time to answer, so said nothing; only they quickened their pace and hurried on. Two other parties went past us in the same manner before we were through our dinner, puffing and blowing like porpoises, but not having time to answer a question.

Such a scramble as this suggested something of interest to men like us who had come thousands of miles, and had gone through with so many hard experiences already in our search for gold, and we grew more and more anxious to know what it all meant.

Presently we saw coming a company of five men almost on a trot, and the foremost one we recognized as the man who held the claim next to ours on Manker Creek. I said to my companions, "Now we will find out something about it." So when he was opposite to us, I sang out, "Hello there, what does all this mean?"

He halted, set down his gun, and was about to answer my question, when his companion came up to him, and giving him a poke with his gun barrel said, "Go on, go on; don't stop a minute."

And picking up his gun, he started off almost on a run; so we were again left in ignorance concerning the cause of the excitement.

Our anxiety was getting the better of us, and we were discussing the propriety of joining in the stampede. But how could we? We had with us only provisions enough for three more meals, and besides, we were nearly worn out with fatigue and exposure. So the matter was settled that we would follow out our first plan ; and after finishing our lunch, we slung on our packs, and started on up the trail.

We had gone but a few hundred yards when we came upon thirty or forty men seated on the ground, who with their packs laid off; seemed to be enjoying a good rest. Sitting down beside them we again asked to know the cause of all this excitement. One of the men, who seemed to be the head of the party, said that a strike had been made upon a small stream on the opposite side of Tonsina Lake, and which emptied into it near its foot, and that the rush was to that place. I asked the man if the discoverers were known to him, and if he thought that confidence could be placed in the report. He informed me that he was the discoverer, and that he and most of his party had claims already staked. This accounted for their leisurely actions while others were so nearly wild about it. I asked if he had any proof of the richness of the strike, and he pulled from his pocket some fine nuggets, which he said were taken from near the surface.

They were indeed a panacea for such cases of sore eyes as affected the most of us just about now,—eyes sore for the sight of real gold in our immediate vicinity. Hadn't we been hunting long and wearily for it, over hills, through swift waters, and amid all sorts of perils? And hadn't it thus far evaded us, with only just enough of the flour dust to keep us eager and hopeful? And the sight of these nuggets, wasn't it a sudden cure for all our tired feelings? Why, we felt at once that we could walk all day and all night, climb mountains, swim or ford streams, and brave perils as we had been doing, if only there was a tolerably sure thing that such finds as these awaited our efforts.

We remained seated upon the ground while the large company put on their packs and started down the trail. We then began to plan if there was any possible way by which we might go to the new strike at once. The question of provisions was the question we were trying to settle. It wouldn't do to be reckless of life in our eagerness to secure a share in this new find. We must take our own supplies along, or not have any. Away up there in those vast wilds were no "grub-stations" where an empty pack might be replenished when once it became empty.

At the camp we had just left, fifteen miles back, were some beans, rice, and dried apples, and if we could live a week upon those three articles, we could go. Could we? That was the question now. The thought of the nuggets of gold was a wonderful argument in the affirmative. Yes, we thought that we could; the thing was decided, and we at once turned back.

Two of us took the pack of the third, dividing it between us, and sent him to the lake camp after provisions, with orders to meet us at the foot of the lake the next day, where we must all stop to build rafts upon which to cross the lake. It was three o'clock when we were ready to start back, and it was twelve miles to the lake.

Two other men came along at this time who, like ourselves, had been on the homeward trip, but meeting the rush, had turned about and joined in, so they kept us company.

We had traveled about three miles when we met a man upon whose face was such a worn-out and haggard expression that I shall never forget it. We asked him where he was bound, and he answered, "To the new strike." I said to him "You are going in the wrong direction," and, added that if he wished to go to the strike, he could follow us. He said that he had been in company with some men, but in some way they had got separated, and he had lost his way. We traveled until eight o'clock, when it became so dark that we could not see to pick our way through the brush. We were yet two miles from the lake, but decided to go into camp until daylight

We looked around for the man that had been lost, but he was nowhere to be seen. Then we hallooed after him, and received an answer from a short distance up the trail. Soon he came along, staggering as he walked, pale as a ghost, and with a look on his countenance to remind one of the contestants in the finish of a six-days' walking match. We were greatly surprised that he would not stop with us, but kept on toward the lake. We said all that was possible to get him to remain for a night of rest, but his only answer was that he must hurry on. We retired early after having prepared and eaten our suppers, but the fear of oversleeping kept us awake much of the night.

At the first appearance of day, we ate our breakfast, and were ready to set out on our tramp. The morning was clear and cold, and a heavy frost covered everything. At the lake we found fully seventy-five men at work building rafts upon which to cross the lake, which at this point was one fourth of a mile wide. One large raft had just landed a dozen men upon the opposite shore, and another smaller one was on its way over.

All the dry trees near the shore had been cut by those in advance of us, so going back seventy-five yards from the shore, we cut three trees, out of which we cut six logs fourteen feet long. Some men who already had their claims secured, kindly helped us to carry our logs to the lake. This done, we were obliged to await the arrival of our companion who had gone after our provisions, and who was also to bring some rope and spikes from the camp.

The whole company that we found at the lake on our arrival had crossed before noon, but we were obliged to wait. The wind began to blow a gale down the lake, and by two o'clock, when we were ready to go over, the waves were so high that we dared not make the attempt to cross. The creek upon which the strike had been made was only two miles from the opposite shore of the lake, and it may be imagined that we waited very uneasily for the wind and waves to become quiet.

Large numbers of men continued to arrive until after dark, when the forest was lighted up by numerous camp fires in all directions.

In the morning there was not a ripple on the waters, and before it was fairly daylight we had our raft loaded, and were out on the lake, paddling for all we were worth for the opposite shore.

Making a landing, we hastily fastened our craft to a bush, shouldered our packs, and started, double-quick, for "Quartz

Creek," as this new strike had been named. On reaching it we found that it had already been staked with claims its entire length, the claims being nine hundred feet square.

At the present time the creek flowed in a direct line to the river, half a mile from the mouth of the canyon; but many years before it had turned around the foot of the mountain, emptying into the river several miles below. The old creek bed was plainly visible, though now covered with timber. Going back four hundred and fifty feet from the stream, we staked two bench claims upon this old creek bed. So many men came flooding in that the whole country for miles around was soon staked.

We wished to satisfy ourselves as to the existence of gold upon this creek, so we went some distance up the canyon, and spent several hours panning out dirt in various places near the surface, and the tests convinced us that at least some parts of the creek contained gold in paying quantities, but how rich yet remained to be proved.

Two weeks prior to this date, two men had been prospecting this creek, and had made the discovery that it contained gold. They at once staked seventeen claims for themselves and the other members of their company. Then they wrote each man's name with a number on a slip of paper, and putting all together in a hat, drew them out one by one, thus locating the claims in such a manner as to give the absent members the same chance with themselves.

Then they returned to camp, pledged to secrecy, until they could advise other personal friends of the find, and let them in "on the ground floor." But one of their number had a friend whom he wished to favor, and he told him of the strike; and this other man also had a friend to whom he communicated his secret, and in this manner the secret leaked out. When the party saw that it was known, they made it public, and the rush began. Men left the camp at all hours of the night, and many started with but one blanket and very little provisions, seeming to forget everything in their eagerness for gold. It was told of one man that when he heard of this find, he just grabbed two biscuits, and ran. Another took with him but a single loaf of bread for a journey of ninety miles, round trip, saying nothing of the time it would take him to locate and stake his claim after he got there.

We saw two men that could scarcely walk, who claimed to have made the entire distance of forty-five miles the day previous, each carrying a pack of fully forty pounds. This might not be regarded as anything wonderful over good roads and in a level country, but under conditions which existed here it was certainly a remarkable feat of physical endurance.

Five miles below Quartz Creek was another stream about the same size, which also emptied into the Tonsina. We thought it possible that this might also contain gold, so we decided to visit it before returning to camp, and about the middle of the afternoon set out through the thick forest which covered the valley from mountain to mountain. That night we camped within a mile of the creek, and in the morning pushed on, reaching it about ten o'clock.

Starting up the stream we soon came to where some one who had preceded us had marked a tree, and written his name, and given notice that he had the day before located this spot as his claim, and that this tree was his north center stake. After describing his claim, and attaching his signature, he added, "I am hungry, and have nothing to eat." One of the boys took a pencil, and wrote under it: "You must be the man who left the rapids with only two biscuits." We remained upon this creek two days, and prospected up for twelve miles; but not finding anything satisfactory, and having only beans and rice remaining in our "grub-sacks," and these without salt, we decided to commence our homeward march, knowing that our home trip would consume three days, and before that time even these would be exhausted.

At dark the first day we reached the lake, only to find that our raft was gone. Some one had crossed upon it, and had left it on the other side. A small one was there, however, belonging to some one else, and on this we crossed, and went into camp.

Some men who had heard of the strike had just arrived, and had come about one hundred miles, hoping to be in time to get claims. Of course they had got left, as there were none for them.

The next morning we were early on our way, and traveled until four o'clock in the afternoon, when we met two of my tent mates coming out to search for us. We had been expected to reach camp four days before, and not having put in an appearance, they became anxious about us, and determined to institute a search, and if possible learn the cause of this delay. They of course were greatly pleased to find us safe and sound, and turned back with us, and three miles farther on we again camped for the night. Resuming our march next morning, we reached our camp at Lake Klutina about 5 P. M.

We were sorry-looking specimens of humanity, I assure you, for our clothing was torn into shreds by so many days' scrambling through such tangles of brush as we had pushed through; but we were heartily welcomed by the boys in camp, who were anxious to learn the results of our trip, and know why we had remained away so long.

CHAPTER XXI

THE TRIP OVER THE GLACIER

AFTER RUBBER BOOTS

SOME months before, we had ordered some hip rubber boots for each man in the company, and in our absence word had been received that they were at our store at Valdez; so it became necessary to send men after them.

At this time of the year going over the glacier was a most dangerous undertaking; in fact, travel had nearly ceased over this treacherous field of ice. But some one had to go, and as was my custom when there was a dangerous piece of work to be done I called for volunteers. Several men quickly responded, offering their services, and it was finally settled that Messrs. Harry E. F. King, Daniel O'Connell, Frank Hoit, and Richard Voigt should be sent.

It was indeed a perilous trip. They might succeed, and come back to us; but they might find a grave in the desolate recesses of the mighty ice mountain. But each of them knew well what the undertaking was. They had traversed this same field under great difficulties, and were willing to go again.

They made the journey to Valdez in four days, on the last of which they traveled nearly the whole twenty-four hours, reaching there thoroughly exhausted.

A good rest was necessary before commencing their return trip, but they were detained there several days more than they had expected on account of a fierce storm that was raging along the coast.

At last a day came which seemed to be propitious for their start, and perhaps I can not do better than to let one of their number, Mr. Harry E. F King, narrate in his own words their experience on their return journey. He said:

"Some months before, we had ordered some hip rubber boots for each man in the company, and in our absence word had been received that they were at our store at Valdez; so it became necessary to send men after them."

Photo from the Joseph Bourke Scrapbook, courtesy of the City of Valdez.

"We left Valdez, and started back over the glacier on Monday, September 19, at four o'clock in the morning. Two days before starting two men came in, and told us that they had been, caught in a blizzard on the summit, and that five feet of snow had fallen, so we knew that a hard trip lay before us. There was no snow on the glacier below the fourth bench, and we made good time until the top of this bench was reached. Here we struck the snow, which somewhat retarded our progress, although it was but a few inches in depth, and we could plainly see the crevices.

"We pushed on to the fifth bench, where we met a party of men coming out, who were homeward bound. They reported five feet of snow on the summit, and also reported that on the opposite side, near the top, they had found a man dead.

"The poor fellow proved to be a man whom we knew. He had started out three days ahead of us, with his partner, and getting caught in the terrible blizzard, had perished. He had died of exhaustion, and the men who went up to secure and remove the dead man's body found his partner insane, and removed him from the glacier.

"As we pushed on up the summit, the snow became deeper and deeper, and the crevices were nearly arched over, so that traveling was exceedingly dangerous. To add to our troubles, each man of us carried on his back a fifty pound pack; so we concluded that caution was no sign of cowardice, and stopping, we took out

"It was fearfully cold, and the wind quickly blew the trail full, after the passage of each man. I kept probing the snow with my alpenstock for crevasses, and found them more frequently than was desirable. In some places we could jump across them, but in many places we were obliged to make a wide detour."

Photo by Neal Benedict, Messer Collection, Courtesy of the Cook Inlet Historical Society.

a three-eighths inch rope about one hundred feet long which we had brought with us for the purpose, and tied one end to my waist, then the next man about thirty feet farther back made it fast to his waist, and so on until we were all fastened together. Thus strung together, we proceeded on over the crest of the summit, and about thirty feet apart, so that in case one of us should fall into a crevice, the others could draw him out.

"It was fearfully cold, and the wind quickly blew the trail full, after the passage of each man. I kept probing the snow with my alpenstock for crevices, and found them more frequently than was desirable. In some places we could jump across them, but in many places we were obliged to make a wide detour. It gave us a great feeling of safety to know that we were bound together by stronger ties than those of friendship, or ties of blood-bound with hempen cords.

"The snow was growing deeper as we advanced, and we sank into it nearly to our knees at every step, and with our heavy packs on our backs we soon became very weary. The thermometer was hovering about zero, and we began to fear that we, like so many others, might become exhausted and not be able to reach our destination.

"The sun went down, and night closed in while we had yet many miles to make before reaching our sleeping bags, which we had left at the foot of the glacier on our way out. We knew that this part of the glacier was well torn up, or rather rent asunder, and contained many yawning chasms. The only thing, however, for us to do was to keep going, so we went on, carefully feeling every foot of the way with our alpenstocks.

"Any one who has not been there can not imagine the absolute solitude of the top of that glacier. Bordered on the east and west by gigantic mountain ranges, and as far as the eye can reach north and south nothing but snow and ice comes within the sweep of vision. And these are cut and seamed at this season of the year by countless crevasses and chasms.

"Not a single living creature within sight or sound of us four, as we trudged slowly and wearily down, till at midnight we reached the lower moraine, a name given to a five-mile stretch of the glacier whose descent is only moderately steep, and begins about three miles from the summit. We were still tied together, and so tired that we could go no more than fifty to seventy-five feet without stopping to rest, but the intense cold would soon drive us to our feet again.

"We kept on until 1: 30 A. M., when we could go no farther. We had been twenty-one and a half hours continuously on the trail, so taking off our packs we made a pile of them, and while three of us kept tramping up and down, the fourth man lay down and slept ten minutes. The extreme cold made it unsafe to sleep longer at a time, but by taking turns, we each obtained a little rest, and at four o'clock we started on again, reaching our sleeping bags at seven in the morning. At nine o'clock we turned in, and slept until six in the evening. Then we got out for supper, and soon after again took to our bags for the night, sleeping soundly until daybreak, when we awoke much refreshed.

"From this point we had to carry our sleeping outfit, which increased our burden to fully seventy pounds to the man,—our pack of rubber boots, it will be remembered, was fifty pounds,— and had two more days of hard travel ahead of us to reach Lake Klutina. But we made it, mentally vowing never to take that trip again at this season of the year."

CHAPTER XXII

PREPARING WINTER QUARTERS

THE weather was gradually getting colder, and the Indians told us that "in two moons, ears break off; fingers break off;" and as it would take us some time to build comfortable quarters, we decided it was time to commence them.

A few campers were making preparations to spend the winter upon the little island, but we knew that during the long months of cold weather chilly winds would sweep up and down the valley, and considered that that place would be too much exposed, and so decided to move.

The larger portion of the campers had already moved down, and were building their cabins along the river. At the head of the rapids, five miles below the foot of the lake, a little village had been started, and fully two hundred people were there preparing places in which to spend the long, cold, and dreary winter.

In the immediate vicinity of the village there seemed to be a scarcity of wood, although plenty could be obtained by carrying it a few hundred yards. But we feared that there might be times before spring when even this little distance might prove a very serious matter.

At the foot of the lake was also a settlement of perhaps a hundred people who were still living in tents, and the larger portion of them were intending to move farther into the interior to spend the winter. About halfway from the foot of the lake to the rapids was a dense forest. Here several of the campers with whom we were the most acquainted had located, and were busy building their winter quarters. Here was an abundance of timber with which to build, and plenty for fuel; and here also we would be much sheltered from driving winds. So we decided to make this spot our home, and selected a site upon which to build our cabin.

"About halfway from the foot of the lake to the rapids was a dense forest. Here several of the campers with whom we were the most acquainted had located, and were busy building their winter quarters. Here was an abundance of timber with which to build, and plenty for fuel; and here also we would be much sheltered from driving winds. So we decided to make this spot our home, and selected a site upon which to build our cabin." Photograph reproduced from the first edition.

As soon as the four men had returned from Valdez with our rubber boots, we broke camp, loaded our boats with goods, and started down the lake. The breeze was so light that we made slow progress, and the first night went into camp only seven miles down from the island. The next morning there was not a ripple on the water, so getting out the oars we put in a hard day's work rowing; but our boats were so heavily laden that we only made twelve miles, and we again landed and passed the second night on shore.

On the morning of the third day there was a little breeze blowing down the lake, and after breakfast we hoisted some sail, and moved slowly on our course, keeping close in shore, so we could land in case a sudden storm arose, as it often does at this season of the year, and was as likely as not to come without giving us any warning of its approach. Several men who had preceded us had not taken this precaution, and had allowed their boats to drift away from shore, when a storm came up suddenly, and before they could make shore, their boats were swamped.

The breeze which favored us in the early morning died away before noon, and we were again obliged to take to the oars. The afternoon was far advanced when we reached the little village at the foot of the lake, but the place selected for our winter camp was two and a half miles down the river from the lake, and the current being swift, we had some fears about our being able to make the landing with our heavily loaded boats where we desired; so we sent two men ahead, to catch our bow-line, when we should throw it, and pull us in shore.

Giving our men half an hour start, we loosed our boats, and pulled out into the stream, and were soon going at an exciting speed down stream. The men had barely reached the landing when our advance boat hove in sight, and throwing them our line, we were speedily pulled in shore. One after another our boats arrived, and were made fast, and soon the entire fleet were resting safely, side by side, on the gravel bar.

Near the landing was a grassy plot upon which no timber grew, and here we set up our tents, and cached our goods as they were unloaded, until such a time as we could build for them a substantial storehouse.

One more boat load of goods remained at the island we had left, and the next morning a large number of men hauled a boat up the swift current of the river to the lake, where luckily a breeze was blowing in the right direction. So hoisting sail, three men swung out up the lake after them, and the others returned to camp, and began in earnest the work of building our cabins and storehouse for our goods for the winter.

CHAPTER XXIII

JIM AND HIS DONKEY

ONE night a member of our company, whom I will call "Jim," on returning from the little village at the lake, informed us that he had been given a donkey, and wished to return next day and bring him down. These animals were valuable in this country, and while some of the boys were congratulating him on his having received such a valuable gift, others declared it was some old crowbait of a beast, too old to eat, and too poor to stand up alone. Jim's reply was that the donkey was not too old to eat, but was all right, and in the very pink of condition, as they would see when he returned with it. He said that he had refused two offers for him already, thinking that if he was valuable to others, he would be to us.

Some of us had been planning another prospecting trip a long way over the mountains, which would be a very hard one, and had been thinking that this donkey might be just the thing for this occasion, to carry some of our luggage. While we were discussing the matter, I noticed a sly wink in Jim's eye, but kept still, deciding to wait until I should see him before making comments.

The next morning, as soon as breakfast was over, Jim took a piece of rope, and started off up the trail. The men went to work cutting timber for our buildings, and Jim and his donkey were soon forgotten.

The day's work was over, and we were all inside our tents eating supper, when suddenly we heard the braying of a mule a short distance up the trail. Talk about a scramble! Coffee was upset, plates of victuals turned upside down, and every one sought to be first outside, to see Jim's donkey —and we all saw it. Donkey ?— Yes. But it was no bigger than a good sized dog, and was just two months old!

We were told—and it was no doubt true—that this was the first young donkey born in this frozen country, and we thought that very likely it might not be long before he would wish that he hadn't been born at all, or it had happened somewhere else. We all prophesied that he was too young to eat grass, and being taken from his mother at such an early age, would not live, but Jim said he could soon teach him to eat, which he proved to us, for in a few days he would drink dishwater, eat pancakes, beans, rice, bread, potatoes, and even meat. In fact, he eagerly devoured all the scraps coming from the table. We have regretted that our kodak fiend was not present, for Jim's donkey and our crowd of admirers would have furnished the very cream of a picture for any comic collection desired.

During the day, when let loose, he would follow the boys about their work like a pet sheep. He also served as watchdog, for any approaching footsteps during the night would be a signal for him to elevate his nose, open his mouth, and send forth upon the still night air such a succession of brays as would awaken every sleeper in camp. Many a man, when thus awakened, would breathe out cruelty toward Jim's pet, but when morning came, would have only feelings of tenderness toward him, and declare that Jim's donkey was all right.

Most of the men in camp had left some tender ties behind them in the home-land from whence they had come, and however rough the exterior might appear, or whatever expressions of thoughtlessness might escape them, still there was hidden away beneath this exterior something very tender; and a young thing like this, which had begun its little life away up here in this desolate country, motherless, and friendless (except for us), and so trusting withal, who could harm him? Who could wonder that within a very short time he was as much the pet of the whole company as he was Jim's?

The site selected upon which to build our winter fortress was back from the river some seventy-five yards. Here the moss covering the ground was six inches in thickness, and so full of mice that they very soon became a source of much annoyance to us. They were industrious workers, for when they were fortunate enough to come across one of our sacks of beans, rice, corn, or pearl barley,

"We decided to build it fourteen by sixteen feet, and set it on posts six feet from the ground, placing on the top of the posts large inverted tin cans so our little "night-thieves" could not climb up them."

Photo reproduced from the first edition.

they would carry loads of it—that is, mice loads—to stow away in any nook or corner which promised a simple and convenient storehouse for them for their winter supplies. Our boots and coat pockets seemed especial favorites for such uses, and it was very common to find from a teacupful to a pint of some of these articles in our boots or pockets.

Their depredations were so great that it was thought best to build our storehouse first, before we did our cabins, for we could sleep under tents with no great inconvenience yet for some time, but we could ill afford to see our stores of provisions, which we had "toted" for so many weary miles, mostly on our backs, being wasted thus by these little intruders.

We decided to build it fourteen by sixteen feet, and set it on posts six feet from the ground, placing on the top of the posts large inverted tin cans so our little "night-thieves" could not climb up them. The floor was made of poles large enough to hold our fifteen tons of provisions. The sides were made of logs, and the top of poles covered with moss, and a thin layer of earth. Before this storehouse was completed we had brought down the remainder of our goods, a complete inventory was taken, and they were stored away for winter.

Our cabins, in which to spend the winter, were the next thing to engage our attention. We planned to build two, sixteen by twenty feet inside, with a fireplace in one end of each. By the first of October these were well under way, at which time I had decided to return to the States rather than spend the long winter in the frozen North.

CHAPTER XXIV

PREPARING TO RETURN TO THE STATES

I INFORMED my friends from Hornellsville, and together we discussed the matter, coming to the conclusion that we would all of us sever our connection from the company, believing we could do much better alone. This was done with many regrets, for the boys were like brothers to us, and our relations with them for all these months had been the most cordial possible.

The matter was brought before the directors, and a satisfactory division of the provisions and other goods effected. Resigning my position as general superintendent, vice-president, and director, I began making preparations for my return home. My goods had to be disposed of and many other things to be done, which kept me busy for several days.

While thus occupied, there came to our camp two brothers, Jesse and Harry Butler, of Clarinda, Iowa, and Charles Barker, of Salamanca, N. Y., who said they had sold out, and were about to return to the States. They had heard that I was going, and thought that we might travel in company. This was gladly agreed to, and October 8 was the day set for our start.

The glacier over which we had come in the spring was now exceeding dangerous to attempt to cross, and hearing that the trip had been made easily in four days by way of Copper River, we decided to go that way.

The distance down the Klutina to the Copper River was thirty-five miles, and the whole distance was over continuous rapids, the current being so swift that it took but one hour and forty minutes to shoot them. But one can easily imagine what a perilous trip this would be. We were told that not more than one out of four boats that made the attempt ever reached its destination in safety. Not caring to take so many chances with our lives and property, we decided to pack our goods to Copper River, and proceed by boat from there.

Two of my friends belonging to the Manhattan Mining Company, of New York, were going down to Copper Center, a little village of about three hundred people situated at the junction of the Klutina and Copper Rivers, and they kindly consented to assist me down with my goods. My companions had engaged a horse to help them down with theirs.

The eighth of October came, the sun shone out beautifully, and the day was perfect. The time had come to say good-by to my friends, I to go, and they to stay. This was no easy thing to do. We had shared a thousand hardships, and been companions in many dangers, and it was like leaving brothers. It was afternoon when I gave the final handshake and set off with my companions down the trail.

CHAPTER XXV

TRIP DOWN THE RIVER

Two nights we camped on the road, and about noon the third day we reached Copper Center. After dinner we went out and found the boat on which we were to make the trip to the coast. It was strong and well made, and capable, as we thought, of standing almost any tests which it might be the fortune of navigation to require of it, and in moderately smooth water had a carrying capacity of about one and one-fourth tons. It had two sets of oar-locks, five oars, and two steering paddles. We loaded into it our effects,—not a very large amount were we to bring back,—and took ship ourselves ready for a start.

A Siwash Indian, who happened to be at Copper Center, but whose home was forty miles down the river, asked the privilege of accompanying us as far as his home; and we were more than glad to grant his request, as the stream was practically unknown to us, while every kink and curve and snag was familiar to him.

About four o'clock in the afternoon we pushed out into the swift current of the river, and were borne rapidly down stream. The day was bright and beautiful, and we looked forward to a pleasant trip, expecting that four days would take us to Orca, where we were to take a ship to Seattle. We had provided ourselves with eight days' rations, allowing for four days of unexpected delay.

When about ten miles down, just as the shades of evening were settling over mountain and stream, we came to an Indian's house, and our Indian pilot informed us that this was a good place to camp for the night. So we ran into a little cove, landed, built our camp fire, and prepared supper.

Several Indian women and children came down, and eagerly

Copper Center with Mt. Drum in the distance. Joseph Bourke photo, courtesy of the Wulff Collection, Valdez Museum and Historical Archives.

watched the food as it was being prepared. We judged that they were hungry, and gave them some, which they ate ravenously; then they returned to their house, and sent a new delegation down after more food. These did not fare so well.

Just as we were finishing our supper a young Indian lad about fourteen years old came up from the river banks to the fire, his clothing dripping with water, and with a broad grin on his countenance gave us to understand that he had just come down the river from Copper Center on a stick canoe (raft) to see his dutchman (girl). We examined his raft. It was made of two sticks seven inches in diameter and seven feet long, tied together, and upon this, with his feet hanging in the water, and the thermometer at zero, he had ridden ten miles down the swift Copper River this dark night just to see his girl.

I wondered how often the young men of the States would go to see their best girls, if it had to be done under such difficulties, and with such a conveyance. I wondered, too, how they would be received if they walked into the presence of their lady love looking as much like a drowned rat as did this young Indian. But love is love,

Copper Center: "After dinner we went out and found the boat on which we were to make the trip to the coast. It was strong and well made, and capable . . . of standing almost any tests which it might be the fortune of navigation to require of it, and in moderately smooth water had a carrying capacity of about one and one-fourth tons."
Photo from the Wulff Collection courtesy of the Valdez Museum and Historical Archives.

whether in the breast of an uncultured savage or a white aristocrat.

After supper we went to return the calls of our dusky visitors. The outside door of the house we found not over three feet in height, and only eighteen inches wide. We did not ring the door-bell, nor did we rap, for a buckskin latch-string was hanging outside, which we gave a vigorous pull, and immediately the door opened. We got down upon all fours, and crawled inside.

The house was built of logs, one story high, and about eighteen by twenty feet in size. It had no windows, light coming from an opening in the center of the roof about six feet square. Upon each side, and running the entire length, was an elevation—a Yankee would call it a wide shelf—about four feet from the ground and about the same in width, and upon these the member of the household, as well as their company, sit, and here we found the occupants now.

The members of the household, in this instance, were fifteen in number, ranging from the gray-haired grandsire of seventy or more to a wee baby of perhaps a month old. This baby was placed in a little wooden tray just the length of its body, and over it was another similar to it, except that this was cut off just below or at

"After supper we went to return the calls of our dusky visitors.The outside door of the house we found not over three feet in height, and only eighteen inches wide. We did not ring the doorbell, nor did we rap, for a buckskin latch-string was hanging outside, which we gave a vigorous pull, and immediately the door opened. We got down upon all fours, and crawled inside." Photograph reproduced from the first edition.

the neck. These were strapped together, and hung up in the center of the room from overhead, and kept swinging by any one of their number who moved about the room.

Their sleeping apartment was a long, narrow wing built upon one side of the main house, and entered from it by a door similar to the other, low and narrow. Their beds were of blankets and the skins of wild animals, arranged in a row upon the ground. In this one room the entire household sleep, no matter how large.

At the rear end of the house was a small door, also, leading into a cave, which is used by the Indians as a bathroom, and in this little underground room they take a sweat bath every day. They build a fire around large stones, heating them very hot. When sufficiently heated, two Indians generally enter, and take their baths together. Pouring water on the stones, a steam arises, which very soon starts a copious sweating, when they begin to switch each other with small switches made of a handful of small boughs of some tree, until the bath is considered finished. This usually lasts about fifteen minutes, and until perspiration is coming from every pore. They then come from their bathroom, rush down to the banks of the Copper River, stop a moment to cross themselves, and take a headlong plunge into its cold waves. After a moment or two of this cold plunge bath, they come out, and often sit on a log, or stand and talk for ten or fifteen minutes, before returning for their clothes. When we were there they stood entirely naked, and talked fully ten minutes, when the weather was so cold that we wore our overcoats, and were cold at that. And this we were told was their daily practice.

It was a mystery to me how those people could come out of that hot place steaming like a locomotive, and while the mercury was at zero or below, go down and plunge into the icy water; and after such a plunge, come out, and sit or stand for so many minutes without a shiver or twinge of muscle.

On first entering their room, we noticed upon our right, sitting in the center of the elevation, an aged Indian more gaudily dressed than any of the others. He had large rings in his ears, and another in his nose, and his clothing, which was made of caribou skins, was profusely decorated with beads, bears' and eagles' claws, and brass trinkets. He beckoned us to him, one by one, and extending his hand, gave each of us a hearty handshake, and motioned us to sit down, thus giving us to understand that we were welcome.

On the opposite side of the room, reclining on a bed of skins, was an old man whose pinched and haggard face betokened suffering. We pointed to this old man, and asked the chief the cause. He replied, "Sick, sick." Then pointing first to his mouth and then to his stomach, he added, "No muck-muck" (food). He meant by this that the old man could not eat.

Then pointing to each finger and thumb of both hands, as if counting them, he raised his hands to his own closed eyes, and said, in a sorrowful tone, "Die, die." By this we understood that in ten "sleeps," or ten days, the old man would die. I crossed over to where he was lying, and took him by the hand. He seemed to have a high fever of some kind. I pointed to him, and asked "Is Siwash sick?" He gave his head a slow nod, and began counting his fingers as the chief had done, and in a mournful voice said, "Siwash die."

I gave him some quinine, and told him how to take it, and tried to impress on his mind that he would not die in ten "sleeps," but it was all in vain. It seemed to be so thoroughly impressed upon his mind that he would die in ten days that I have little doubt but that he did die in the appointed time.

A young Indian came over to where I was standing by the sick man, and in the few words of English at his command, told me how the old man would die in ten days; and then motioned how a grave would be dug, the old man lowered into it and covered up with earth,—all in the old man's presence. It must have been sorry comfort for the poor old man.

Across the room, in one corner, sat the young Indian who had come down the river on his stick canoe to see his girl. Beside him sat a young girl, perhaps ten years old, the one for whom he had braved the cold waters, coming ten miles with his feet hanging in the cold Copper River. He would look at us, and then at the girl, and giggle. She would look at us, and back at him, and giggle. We had little doubt but that they understood each other, though not a word was spoken.

We remained there about an hour, when we returned to our camp, crawled into our sleeping bags, and were soon away in dreamland.

The next morning dawned clear and cold. The little cove into which we had pulled our boat was frozen over, and it was fast in ice. Axes were brought, and soon an opening was chopped out into the main stream. Breakfast over, our camp outfit was soon loaded in, and we were again floating down toward civilization.

We soon learned that our Siwash friend, who was our guide and pilot, was the chief medicine man of the whole tribe, and he must stop at every Indian house along the route to ascertain if any sick were there, which frequently was the case. This gave us an excellent opportunity of seeing the Indians in their homes.

During the day we came to a small Indian village. A few cabins were clustered together around a large one, which was the home of the chief, Nickoli.

We followed our guide into one of the cabins, where he gave us to understand that the chief's sister was sick. We found two old women sitting upon the ground floor of the cabin near a fire; one of them being the sick woman referred to. She was a mere skeleton; her face was haggard and wrinkled, and she looked as if she might have begun life about with the century.

The medicine man spoke a few words to her in the Indian tongue, unintelligible, of course, to us; then going to the side of the cabin, took a cup, and taking a small package from his pocket, which contained some kind of leaves, emptied a small amount into the cup, upon which he poured a little warm water. He then went behind the sick woman, as she sat upon the ground, and clasping both his hands around the cup, rested the bottom on the bowed head of the patient. Then looking upward, and puckering up his mouth as if to whistle, he gave two or three hard blows, as if to blow the disease away. He then let go with one hand, passed it back and forth several times over the cup, and raising it heavenward, gave several beckoning motions, as if calling a blessing upon the medicine. Then he looked down toward the cup, at the same time placing his hand over the top of it. For a moment he remained motionless, then handing the cup to the sick woman, she slowly drank its contents. Handing the cup back to the doctor, he went through about the same ceremony, and set the cup back upon the shelf.

Seeing that he had done all for the sick woman that he intended to do, we reminded him that we must be going; so crawling back through the low door by which we had entered, we returned to our boat, stepped into it, pushed out, and soon the little Indian village was far out of sight.

All day long Mounts Sanford, Drum, and Blackburn, with their snowcapped peaks, as well as the dome-shaped top of Mount Tillman, out of which were still rising volumes of smoke, could be plainly seen.

During the afternoon, while we were stopping at an Indian cabin, a boat nearly the size of our own, containing four men, passed us on its way down the river. The afternoon was far advanced when we came in sight of the home of our Siwash guide. Just before reaching his cabin we saw the party who had recently passed us, having gone into camp for the night but a few hundred yards from his house.

Our guide had invited us to camp near his place overnight, so

bringing our boat ashore, we unloaded our outfit, and began looking for a suitable place. There was no wood near for our camp fire,
so our Indian friend went up to his cabin, and soon two Indian
women came down to us with large armfuls of wood, ready cut.

We noticed several large silver salmon hanging upon a rack
near the river, where they had been recently caught. We decided to
have one of them for our supper, if possible, so we made our
wants known to a small Indian boy standing near, and he ran
quickly up to the cabin, and soon an old woman came down to
make the sale. We traded two cups of beans for an eight-pound
salmon, and for a six-pound fish gave one cup of rice.

After supper we visited the Indians, and found seventeen of
them occupying the same cabin. We had been seated but a few
minutes when an Indian lad of about twelve years, wishing to
show off his musical ability, went and brought out an accordion,
and favored us with selections. We were not a little surprised to
see this untutored lad show so much skill in the use of any musical instrument. From some of the white men he had learned a few
lines of "Marching through Georgia," "John Brown's Body," and
"There'll Be a Hot Time on the Old Town Tonight." These he took
great delight in singing. The tunes he rendered very well, but the
words were a little too much for his thick tongue to manage, and
make very intelligible. The boy began, and one by one the others
would join in, until in a little while the whole company were
singing to us.

After an hour spent in song, in which they favored us with
several of their own songs, they began to bring out articles for
trade. We bought several articles of their own manufacture, to
bring home as curios. After another hour, we returned to our
camp, only to find that the Indians' dogs had invaded our camp,
and stolen our fish and a large piece of bacon. But we cared little
for that, as we supposed we would have an abundance of provisions for our trip.

We had heard that Wood's canyon , on the Copper River, was
a very dangerous place to pass. The rocks on either side, for a distance of three miles down the stream, rise almost perpendicularly
to a height of several thousand feet, and the channel has an aver-

age width of about one hundred feet. Through this narrow divide all the water of this great river must flow. We had been dreading this place, and here we learned that it was only fifteen miles distant. So we determined that in the morning, if possible, we would secure the services of the same Indian that had piloted us the last forty miles to pilot us through this passage.

The night was clear and cold, and the temperature must have been far below zero, for in the morning the river was full of floating anchor ice. While we were at breakfast, the boat containing the four men who had camped near us overnight, passed us on its way down the stream. The Indian agreed to pilot us through the canyon for two dollars, which we were glad to give, rather than take our chance with the rocks in this unknown channel; and loading up our boat we pushed out into the stream. The day was beautiful, and as the hours advanced, the rays of the sun, shining upon the frosted mountain peaks, painted a picture not soon to be forgotten.

Down this portion of the river the current was not swift, and we drifted leisurely along, feasting our eyes on the beautiful scenery lining its course, little thinking it would be so many days before we should reach the coast.

About noon we came to the entrance to the canyon. Here we overtook the four men who had passed us four hours before. We landed, and made their acquaintance. They were D. T. Peters, of Chillicothe, Mo.; August Winstrom, of Philadelphia, Pa.; Christopher Traveland, of Eureka, Cal.; and Joseph Lawson, of Norwalk, Conn. The latter, it will be remembered, belonged to the same mining company as myself, but withdrew from the company early in the season, before going over the glacier. These men had sold out, and, like ourselves, were going back to the States.

While making the acquaintance of these men, we saw another very small boat coming down the river, containing but two men. Seeing us on shore, they landed also. They were G. H. Winters, of Indianapolis, Ind., and J. Stodart, of Osceola Mills, Pa. Our little company now numbered ten men and three boats.

Here we all prepared and ate our dinner, wondering the while if any accident would happen to us while making the passage of this dreaded canyon.

Soon we were in our boats again, and out among the floating ice. All was excitement on board these three boats as we entered the narrow divide. Every one was fearing a rough time, yet hoping to make the run in safety.

But imagine our surprise when we found the waters moving no more than two miles an hour, and as smooth as a lake. Being so happily disappointed about these reported dangerous rapids, we began to hope that the great Copper River rapids might not prove to be as dangerous as they had been pictured to us.

Having passed these "rapids" our guide left us, he having to climb the mountains forming the great sides of the canyon, on his return trip. It was with many genuine regrets that we parted with our Indian guide, for with him at the helm we felt at perfect ease, knowing that every rock, current, and bar was familiar to him. We tried to engage him to go farther with us, but he would not; so from this point we were our own guides.

Having taken leave of our dusky friend, we pushed out again into the stream. When we came where the bed of the river was rocky, we would row hard down the stream, to give the boat steerage way, so the man at the rudder could steer us around the rocks and rough places. Then when we came to smooth water, we would rest at our oars, and drift with the current, thus giving ourselves an excellent opportunity of viewing the bewitching mountain scenery through which we were passing.

Word pictures may convey to some minds a faint conception, perhaps, of such views as these, but not my words. One must have a rare gift for such delineation, and even then he must feel, after his best efforts, that his work has largely failed to impress others as he himself is impressed.

Gazing upward at these massive walls of wonderful masonry, which have defied the power of ages untold to crumble them at their tops or loosen them at their foundations, one's thoughts are prone to run backward to the time when they were reared without hands, from materials gathered from no mortal knows where.

The three boats ran along side by side where the waters were sufficiently smooth, and we chatted gaily about the incidents of the trip, the beauties of the scenery, and the wild and rugged nature of the country through which we were passing. Occasionally as we rounded some large point of rock, there would open up before us pictures so different from any we had yet seen that we would sometimes get so absorbed in the grandeur around us that we would forget to watch, until the grinding of our boat's bottom on the gravel would bring us back to our senses, only to find that we were stuck fast in a sand bar. Sometimes we could easily push ourselves off, though at other times we were obliged to get out into the water, which would so lighten the boats that we could shove them into deep water, and spring in.

The afternoon sun had long since disappeared behind the great mountains, and the weather being piercing cold we began early to look about for a place to camp for the night. Two things were essential,—a place well sheltered from the wind, and a good supply of wood.

After passing through the canyon, the spruce timber which had served us so well all through the summer, had entirely disappeared. Our only supply now was what floodwood could be gathered along the river, or the green alders fringing the banks.

Upon our right we noticed a small canyon, down which flowed a small, clear stream. Near where this emptied into the river was a little cove and a high, level sandbar, and near this was a large pile of floodwood. This seemed to us an ideal place to camp, so we pulled into the little cove, and made our boats fast, unloaded our camp outfits, and prepared our evening meal. Our good supply of firewood enabled us to build a roaring camp fire, around which we sat, telling stories and chatting, until nine o'clock, when we spread out our beds on the frozen sandbar, with the great blue canopy for our tent covering, and crawled in, and were not long in getting away to the land of nod.

When I awoke in the morning the roaring sound which greeted my ears told only too plainly that the wind was blowing a gale, and upon getting out of bed, I found everything covered with sand and dust and the wind sweeping down the valley at the rate of fifty miles an hour. The atmosphere was so filled with sand and dust that one could see only a few rods, and the river seemed a body of moving ice.

Soon my companions were out, and we discussed the situation. How we regretted not having made better time the day previous, which we might easily have done, and now have been much farther down the river and of course nearer the coast.

However, we determined to spend no more time on scenery, but would get down the river as fast as possible. Hurriedly we prepared and ate our breakfast, and were soon on our way down the river, pulling through the floating ice.

How different from the day before: then the sun was shining bright and clear, and we could see far ahead, and could thus shun many of the rough places; today the weather was extremely cold. Often the river would divide into small channels, and we could not see which to take. True, the wind was blowing down stream, thus helping us on our course; but when we wished to avoid some difficult or dangerous place, it would drive us directly toward it, and the floating ice made it difficult to go anywhere but with the current.

Sometimes we would mistake a side channel for the main one, and when too late to return, find ourselves aground upon some riffle. We had then to get out and push our boats before us until we had sufficient water to float them. Of course this was delightful exercise in zero temperature, but it had to be done. In this way we would sometimes be cut off from the main channel for miles, and not know on which side until we came into it again. This made our progress slow, and gave us some forebodings, for our stock of provisions was getting threateningly low. We had used up over half of them, and had not covered one third the distance to the coast. We could not get back, and should the river get closed entirely, we would be left, as far as we knew, with but few white men or any amount of provisions within a hundred miles of us. We dared not dwell upon the dark side of this picture, but kept hoping for the best.

We had approached a place where the valley widens, and the river, in high water, is eight miles wide; but now, in low water, it was cut up into many channels. Across this wide valley the wind swept the sand like smoke from a burning fallow. We did not stop to get dinner, but pushed on as fast as possible.

How far we were from the mountains on either side we could not tell. Night was fast approaching, and where were we to camp?

Not a stick of wood had we seen for a long way bank; nothing but sand everywhere.

Soon we found we had missed the main river and were going down a side channel. This began to broaden out, and there was no depth of water. Our boat struck against the ice; the stream was frozen over. We could not get back; we must go on. We tried to push our boat along, but the whole bottom was a bed of quicksand, which would prevent us from getting out, should our boat go aground.

For a moment the sand in the air seemed to grow less, and we thought we could discern open water a hundred yards farther down; so with ax and oar we began to break the ice in front of the boat, but it began to grow dark before the open water was reached. The prospect seemed good for us to pass the night without either fire or supper, and we had had no dinner, either.

Coming out into open water we pulled on down stream as fast as possible, hoping soon to come to the main river. In a few moments two sights met our eyes which caused us to feel greatly relieved; a few rods ahead of us was the main channel, and upon the bank was a large pile of floodwood. Here was the place to camp.

We landed, drawing our boats clear up on shore to keep them from being frozen fast in the ice by morning. The wind blew such a gale that it was no easy task to start a fire, but persistent effort brought success, and before long the sound of sputtering bacon was heard, with flapjacks next in order, but while they were being baked, the flying sand would light on them, so that before they were ready to turn the top side would be the brownest of the two. Our party may forget many of the incidents of their stay in Alaska, but I doubt if any of them will forget the eating of those flapjacks and bacon and sand.

The wind blew so hard all night that our sleep was much disturbed, and when daylight came we were half buried in the sand, and had the night been long enough, our beds might easily have become our graves. Crawling out, we prepared breakfast under the same difficulties that we had in getting our supper, the wind still prevailing, and dust and sand still flying.

While eating, we were surprised to see a strange man approaching our camp. We learned from him that he was one of the party who had been all summer pulling their goods up Copper River, and were on their way to Bremner River, which he said was but a few miles above where we then were, where they expected to go into winter quarters. They had been caught in this sand storm, the same as we had, and were obliged to camp on the same island—for it seemed we were now on an island. He told us of another party who were on their way up the river, and were camped several miles farther down; and he also gave us some information in regard to the river which proved of great value to us.

After breakfast we again set out down the river. Everything went well, and at one o'clock we came to a place where the river ran close to the mountain, and there upon the bank, in a small cluster of trees, stood three tents. This was a welcome sight to us, but it was very difficult to make a landing because of the river being so blocked with ice. The men from the tents, however, seeing us approaching, came down to the bank, and caught our bow-line as we threw it, and helped us pull in shore.

These men were also waiting for the sand storm to cease, so they could resume their journey up-stream. They hoped to reach Bremner River, where they said a large party had gone into winter quarters. They also gave us much valuable information about the river, the great rapids, and the glacier, all of which we had yet to pass.

Near their camp was a small sheltered spot where we built a fire and prepared dinner. Here we decided to remain until morning, hoping by that time the terrible sand storm would be over. Our eyes were becoming much inflamed, and should the storm continue many more days, we feared serious results in that direction.

All the afternoon the wind swept the sand down the valley with unceasing fury, and all night long the roaring of the wind through the tree tops and along the mountain side told us that the storm was still on.

Morning came, but brought no change, except that the wind had increased rather than diminished, and the river seemed more blocked with ice. We dare not go out into it while the wind blew such a gale, so we waited all day; but the day passed, and night came on, and still the storm raged on.

The next morning at dawn we were up and out and preparing to move on. The wind had ceased, and the dust had settled, but a dense fog was filling the valley, which prevented us from seeing any distance ahead.

Breakfast over, we again set out down the river, which was so blocked with floating ice that we could go no faster than the current. We drifted on until about noon, when we met some more men, who were pulling their boats up the river. They were almost discouraged, for it was slow, hard work to pull against so much ice.

They told us that it was about thirty miles to the large rapids below, but about five miles below were small rapids, and they advised us to rope our boats down them, as they considered it unsafe to shoot them. We had heard nothing of these rapids before, but it caused us to proceed with caution.

About two o'clock we came in sight of what we supposed was the smaller rapids, and made an attempt to land, but it was with great difficulty that we got on shore, on account of the ice in shore. The place of landing was near the foot of a high mountain. Just below us was a large rock perhaps fifty feet high, which projected out into the stream, and the whole current of the river set in close to the bank just above this rock. and the great mass of ice which filled the stream from shore to shore came rushing along at the rate of ten miles an hour, striking the shore and floating down until it reached this rock, where it was thrown out into the river again to meet the ice coming down on the opposite side, thus forming a current or swirl of water and ice which was most dangerous to pass with small boats like ours.

The bed of the stream was evidently filled with rocks, as the waters, when not held down by ice, would foam and boil in a fearful manner. Making our boat fast, we went down to investigate. Standing upon this rock, and looking down into the mad rush of ice and water, knowing that we had to pass it, was a sight sufficient to cause the stoutest heart to quail.

To get our boats over this almost perpendicular wall of rock was impossible; and to cross to the opposite side was a thing not to be thought of. What to do we didn't know. It had commenced to get warmer, so we decided to go into camp, and wait another day's appearance, thinking the flow of ice might be less. We found sufficient floodwood for our present use, so built a large camp fire, and discussed the situation around its cheerful glare.

It was evident that the little boat which had brought Winters and Stodart safely down thus far would never pass in safety the place above described, so it was decided that they should abandon the little craft, and make the balance of the trip in our boat. Accordingly their goods were transferred from their boat to ours. About this time it began snowing, and continued until after dark. The snow was wet and heavy, melting upon our clothing, so that in a little time we were wet through. Unpleasant as this was, we preferred it to our experience with the sand, which had been such a source of discomfort to us for several days past.

When bedtime came, we spread out our beds upon the clean, new snow, and in our wet clothing, crawled in for a night's rest. To the reader this may not seem an inviting place to spend the night, but with us it was a case of that or nothing. The open canopy had been our bedchamber about all the time for several months. Sometimes, while in a permanent camp, we had occupied tents.

When we arose at daybreak our spirits sank, as we noticed that instead of there being less ice, the river was densely packed. Looking to where our boats were fastened, we noticed that the small one, which had been unloaded the night previous, was gone, having been torn loose by the ice, and taken down the stream. You may be sure we were thankful that it was that, and not one of our larger ones. Had it been one of these, our case would have been a desperate one indeed.

Here, as in many times in the past, I felt that the eye which never sleeps had been watching over us, and a hand mightier than ours had been guiding our course.

CHAPTER XXVI

A THRILLING EXPERIENCE

THERE seemed to be but one way out for us; we must go on, and attempt to pass the rock. We laid our plans, and proceeded to carry them out. It was to work our boats farther up the stream, watch our chance when there seemed to be less ice, and with four men at the oars, force our boats as far out into the stream as possible before reaching the dreaded spot three hundred yards farther down.

A suitable place for starting was reached, and we all stepped into the boats, halting a moment for a favorable chance to push off. Soon it came, and my boat, which was to lead, started, followed closely by the other. With long, steady strokes we pulled with all our might.

The ice through which we were compelled to push our boats greatly retarded our progress, so much so that we were being carried down toward the rock faster than we were leaving the shore. Still we pulled on, hoping to get far enough out to clear the great roll of ice which was tumbling outward as the current piled it up against this great rock.

There was a small clear space of water ahead of us. Our steersman saw it, and told us to pull hard for it. The request was scarcely needed, for every man was pulling his best. Reader, did you ever pull for life? Then you know something of the feelings which nerved every arm on that dreadful trip. We did pull.

Fifty yards above the great rock we came out into the open water that we had been pulling for. This enabled us to shoot ahead faster, and to get farther out into the stream. Here the distance of only a few feet might be the difference between life and death. Ah, a single foot might mean all that. So every man put all his remaining strength into every stroke. I looked at our steersman. His eyes were riveted on the rolling mass of ice and water which we were fast approaching. A moment more, and we should know.

Suddenly he threw the rudder to one side, and the bow of the boat turned down stream; at the same time we were knocked to one side by the ice, and our boat turned half way round, but we were past the rock.

Next our attention was turned to our companions. Would they get through as well? We held our boat back against the current, to be near in case they might meet with any accident and need help, but we soon had the satisfaction of seeing them shoot through as safely as we had done.

We congratulated ourselves, as we drifted leisurely along, on having passed the small rapids in safety, when upon looking ahead, we discovered that we were just approaching them. No possible chance to land for a reconnoiter; we must go through them, and do it right away.

We began rowing down stream as hard as we could, to give steerage way to our little crafts. In a moment we were shooting down the rapids at the rate of twenty miles an hour. The waves broke over our bow at nearly every boat's length. But we soon passed the danger point, and were drifting with the current.

Our provisions were getting low, but we expected within two days at most to be in Alganak, which is the first Indian trading post coming down Copper River.

Nothing worthy of note occurred until after one o'clock, when we discovered a small Indian canoe a short distance below us fastened to the shore. Thinking there must be Indians near by, we determined to make a landing, and, if possible, get some provisions.

The current was swift, and the river filled with ice, so that landing was no easy matter. We kept our boat as near the shore as we thought safe, and watched for a convenient place where we could throw the bow near enough to allow one of our number to leap ashore with a line. Soon such a time came, and as the line tightened, our boat swung in against the bank, and we were on shore in an instant. We called to our friends in the other boat to throw us their bow line, which they did, and we were all landed, and our boats drawn alongside.

Across the flats, about a fourth of a mile distant, we saw an Indian approaching, accompanied by a little lad of probably ten years, barefooted, and trotting along through three inches of new-fallen snow. When he came up to us, we asked him if he had any "muck-muck" (provision). He answered, "Hi, you, muck-muck," meaning "plenty."

Pointing to ourselves, we said, "Hi low, muck-muck," which in their language means "none." Not seeming quite satisfied to take our word for it, he got into our boats and looked through our goods, but not finding any provisions, came on shore, and motioned us to follow him. We began to make our boats fast where they were, but he shook his head, saying in very poor English, "Water steal canoe," and pointed us to a little cove a short distance below, where they would be safe.

When we had cared for our boats, we followed him to his tent about half a mile distant. His wife put her head out of the tent as we approached, and, with a frightened look, said a few words to him which we could not understand. Then he motioned us to come in, and gave us seats on the ground around the side of the tent, and began getting us some dinner.

They baked bread, cooked some evaporated potatoes, and oatmeal, fried bacon, and made some coffee. Our appetites were keen, and the meal was greatly relished by us all.

After dinner we asked him for the bill, but he gave us to understand that it would be nothing. However, we each gave him twenty-five cents, which seemed to please him greatly, for he went out and brought in a half sack of flour, and gave it to us, but would not sell any. He also gave us to understand that two Boston men, as they call the whites, were camped on the opposite side of the river. This made us anxious to cross over, and see if we could obtain any supplies of them, and also gain any information concerning the great rapids below.

We did not suppose that we were anywhere near them yet, but the Indian told us that they began just around a rocky point but two hundred yards away. How fortunate it was that we stopped where we did, for had we gone on, our doom would undoubtedly have been sealed.

The river at this point was two hundred yards wide, with a current of at least ten miles an hour, and was filled with floating ice. To cross where we were without being drawn into the rapids was impossible. Our only chance to get over was to draw our boats back up stream a quarter of a mile, and then pull as hard as possible for the opposite shore while the current was taking us down. This we knew was a dangerous task, for should we fail to make the other shore in time, nothing could prevent our going through the rapids. But our case was beginning to be a desperate one, and we decided to chance it. So hauling our boats up the river until we considered it safe to start, we bade our Indian friend adieu, and pushed out from shore.

It was unnecessary to tell any one to pull his best, for every man knew what was at stake, and what it would mean to fail. We succeeded in crossing the main current, and entered the lower side of a small cove just above the rapids. Here our boats all became fast in the ice, and began to circle around the cove, now being carried within a few yards of shore, then around to within a little way of the swift current leading to the rapids, but to do our best, we could not make the bank.

How this might have ended we will never know, as great odds were against our breaking our way through this mass of swirling ice packed so closely together, but a man on the opposite side saw our predicament, and running to the bank, shouted, "Throw me your line." This we quickly did, and were helped to shore. We were glad to set foot again on shore, for we could plainly hear the roar of the rapids during our swirling boat ride.

Our new-found friend invited us to go with him where he and another man were camped. They had come to Alaska early in the spring, and had decided to go to the interior by the way of Copper River. Coming up as far as Alganak, they had staked some quartz claims, which promised well; but still having a desire to visit the interior, they had brought their goods on up the river, until, after suffering hardships and privations, they had sold out their goods, and were on their way back to their claims.

They had a 12 x 14 tent, and a stove, the use of which they kindly offered us on which to prepare our meals, an offer we gladly accepted. We found the shelter of a tent very grateful, for it had begun to sleet, and was very disagreeable outside.

We were able to buy of these men a good supply of beans and a small amount of other provisions.

CHAPTER XXVII

A "LIVE" GLACIER

THEY told us it would be impossible to rope our boats down the rapids until the snow was melted off the ground and rocks, because the banks were steep, and if slippery, it would be impossible to hold our boats back in the swift current.

A boat containing two men was seen coming down the river about four o'clock, and we ran down to the bank and motioned them to land where we were, which they did with much difficulty.

Like ourselves, they were getting out of the country. They had left Copper Center the day after we had, and of course had experienced about the same kind of weather, and other vicissitudes as ourselves. Their boat being a small one, it was arranged that it should be abandoned at this point, and they make the balance of the trip in our boat.

One of these men had gone into Alaska by way of Skagway, reaching Dawson in the early part of summer, and he with his company had made their way over to Forty Mile Creek, where they found good prospects. Leaving his partners there, he passed over the Tanana Range, to the head waters of the Copper River. There he fell in company with a man who had gone up Copper River, and was returning. Together they came down the stream, reaching Copper Center early in October. Remaining there a few days, they had followed us down, overtaking us as before described.

We were anxious to proceed, but our new-made friends, who were more familiar with the river than we, told us as others had told us before, that it would be impossible to rope our boats down the rapids while the rocks where so covered with ice as then. That night we all tried to sleep inside the one tent, but were so crowded that we got but little rest. Next morning brought no change in the weather, as we had hoped.

"I had seen glaciers before, many of them, and very beautiful sights they had afforded, but Miles's glacier struck me as far surpassing them all." Crary Collection. B62.1.778. The Anchorage Museum of History and Art.

That day we walked six miles over the glacier moraine, to get a look at Miles's glacier. Arriving within a mile of the glacier, we came to a high point in the moraine, which gave us a commanding view of the whole glacier. Alaska is full of surprises; a walk of only a mile sometimes will so transform everything in sight, that it looks almost like another country. We had been meeting these experiences for many months; but from our seat on this elevation I am free to say that I had never witnessed anything so grand. I had seen glaciers before, many of them, and very beautiful sights they had afforded, but Miles's glacier struck me as far surpassing them all.

It had a frontage along the river of three miles, and a perpendicular wall of ice rising out of the water three hundred feet high, and extending back into the mountains far enough to give an area of eighty square miles. Its entire surface was covered with spires, and domes, and peaks, and arches. The summer suns of many years had melted away the soft spots, and left the surface in every conceivable shape.

This immense body of ice—of over eighty square miles—is what is called a "live" glacier, and is said to crawl down toward the river seven inches every twenty-four hours. Great masses of this ice fall into the river every day. Scarcely any five minutes of any day during the summer passes without some of it falling. As the weather gets colder, the falling becomes less frequent, and during the extreme weather of winter it ceases altogether. Two large pieces fell while we sat watching it, making a crashing sound which could be heard many miles away.

We were told that during the summer, when the salmon were running up the river, that sometimes great masses of this glacier fall into the water with such a swash, that wagon loads of salmon are thrown out upon the banks by the waves.

Across upon the right but four miles farther down, yet in plain view from where we sat, lay Child's glacier. It extends down the river three miles, and running back, it circles completely around one mountain peak, which towered far above it like a gigantic sentinel. It covered the same number of square miles as Miles's glacier, but is not a "live" glacier, as it does not move, nor does it crumble off and fall into the river as does the other.

The air between these two glaciers was piercing cold, and we were soon reminded that we must move on or freeze. Our forenoon's explorations and sight-seeing had been long enough and had climbing enough in them to give us a relish for our dinner when we returned.

After dinner some one discovered an empty boat coming down stream, and we all ran out to see what we might discern as to its owner's fate, but it did not come very near in shore, and was soon caught in the rapids and whirled beyond our sight around the rocks.

We wondered where its owners were; had they been capsized and lost, or had their boat been stolen from its fastenings by the treacherous waters and ice, leaving them in some desolate spot far from human habitation, or any means of obtaining food? These were our speculations, but our queries were never answered.

The tragedies of Alaska, their number will never be known, their awful history will never be written. Far away friends will wait, and hope, in blissful ignorance of the terrific struggles and unequal conflicts with nature's giant forces, until they went down, to lie in ceaseless solitude, or be buried without hands under mountains of snow and ice, or washed into unknown waters, or crushed into shapeless mass, or drifted into infinite tangles of floodwood along some of these rushing streams; they will only know that they went away and came not back.

Toward night the weather became warmer, and it began to rain, and we hoped by morning the ice would be so far gone that we would be able to proceed on our way, roping our boats down the rapids.

Our tent leaked so badly that by bedtime, the ground under it was a veritable mud puddle. This must be our bedchamber, or we must sit up; I chose the latter, and passed a very uncomfortable night.

When the morning appeared, the ice was gone from the rocks, and the snow was nearly gone. As quickly as possible we prepared and ate our breakfast.

CHAPTER XXVIII

ROPING THE RAPIDS

WE thought to have passed the rapids by night and to be far down the river below the glacier. So loading our goods into our boats, and leaving the tent and stove behind which did not belong to the two men, we began roping our boats down the rapids.

We had 150 feet of rope attached to each end of each boat, and several men to a boat. The work was slow. The banks were steep, and many of the rocks over which we had to climb were ten to fifteen feet high.

Often as the boats drifted down the current, our ropes would get caught under some rock, and it was with great difficulty that they could be removed.

We frequently came to places where we dared not venture to leave our goods within the boats, lest they should be capsized and the goods lost, so they were unloaded on the bank, and the boats held back until the full length of the line was carried down stream, when the boats would be pushed out into the current to shoot through. If they capsized, we lost nothing, which happened more than once.

We had, of course, to pack our goods over these stretches, and then reload them, and go forward, until another bad place was reached, when the same thing would be repeated.

Sometimes it was necessary, in order to clear some projecting rock, for one or two men to ride down in the boats. This was no desirable job, but some one had to do it. Many men have attempted to "shoot" these rapids, but few have lived to relate their experiences during its passage.

The river here is confined to a width of one hundred and fifty feet, and for three miles its immense volume of water rushed down at a tremendous rate; besides, the whole bed of the stream is filled with rocks, over and around which the water lashes itself into foam, with a noise almost equal to the roar of the great Niagara.

We had the pleasure of meeting one man who made this hazardous trip in an open rowboat, and who came through unharmed. But what made the man a more interesting character to us was the fact that he had gone through just for the sake of the adventure, carrying with him an Indian, who was an unwilling passenger.

He was a man of quiet, unassuming appearance, of perhaps twenty-five years of age. His hair was extremely red, and he gave us the impression that his life had mostly been spent upon a farm. One day while on his way up the river, and while camped near the rapids, he decided to cover himself with glory by shooting these wonderful rapids. There was an Indian living near, belonging to the Siwash tribe, who for stealing had been banished to a desolate spot near the rapids. Here he, with his family, must live for five years, and must catch and dry every year a certain amount of fish for the use of the tribe, but was not allowed to return until the five years were up. This is the Indian's punishment for thieving.

One day this fellow and his companions hired this Indian to help them rope a boat down the rapids. The man and the Indian were in the boat, holding it off the rocks, while his companions were holding the rope. It seemed to have been understood between them and him, for just as they were entering the rapids, he took out his watch, looked at the time, then severing the line which held them to the shore, the boat shot down the rapids.

The man seated himself at the oars and began to row. As soon as the Indian saw that they were turned loose, we threw himself into the bottom of the boat, and began to cry to the water gods to save him. They passed the rapids in safety, and the trip of three miles was said to have been made in four and one-half minutes.

We worked hard the entire day, and when night came had only made two miles. The shore along which we then were was steep, and almost one mass of rocks, which had fallen down from the mountain side.

We tried to reach a more suitable place to camp, but darkness came on, and we were obliged to stop.

We succeeded in getting a camp fire started, and prepared supper, but we could find no place to spread our beds. Some of the men tried leaning up against something, to get a little rest, but the greater portion of us were awake, and kept our fires going all night.

We prepared breakfast and had it eaten before daylight. At the first appearance of dawn we were ready to start. The weather had turned colder, and the river was still full of floating ice. About ten o'clock we had reached the last part of the rapids. For about three hundred yards it was impossible to rope our boats down the main channel, but there was a side channel down which we might go, by removing our goods and carrying them, as at several other places, sending the boats down empty.

The stream was so filled with rocks that we could not even float our boats here, without getting out into the water and pushing them off the rocks frequently. Often we had to wade in the icy cold water to our middles, and by the time we had reached smooth water we were thoroughly benumbed with cold. We did not stop to build fires and dry our clothing, but ran races up and down the beach to get warm, thinking that our troubles were now over, and that we should soon reach Alganak, where we hoped to get a fresh supply of provision for the balance of our trip.

We loaded up our goods again, and being below the rapids, got in and pulled down toward the glacier. But we were barely started when we espied two men on the opposite bank, who ran down to the water, and motioned frantically for us to go back, and we knew that something was wrong. So we effected a landing, and fastening our boats, went down to investigate.

Going close to the glacier we saw a sight which almost caused our hearts to stand still. The whole river, for four miles down, was blocked with solid ice.

CHAPTER XXIX

FOUR MILES OF SOLID ICE

I SHALL never forget the discouraged looks which spread over the faces of the men, as the situation dawned clearly upon them,—men who had braved almost every kind of danger, and borne almost all sorts of hardships. True, we had one hundred pounds of beans left, but one more meal would exhaust every other article of food in our possession. Of course the beans would keep us from starvation, but how long we might be kept here we could not guess.

The outlook was, to say the least, anything but a cheerful one, and to add to our discomfort, the weather was growing colder, and the wind began to blow a gale down the river, bringing with it a cloud of dust. Returning to the boats, we gathered up some floodwood, made a fire to dry our clothes, and began in earnest to discuss the situation,.

The great blocks of ice which had broken off the glacier had floated down here and become massed in the stream, and then the floating ice in the stream, formed by the freezing weather, had also filled into this mass, until there was a solid block four miles long.

We spent the night there hoping that the wind and water together might loosen this great ice jam by morning. We had gone without our dinners, and toward night had cooked and eaten the little food remaining, aside from our one sack of beans. Having had but little rest for two nights past, we spread our sleeping bags down upon the gravel and slept soundly until morning.

At day dawn some of us went down to see if the night had brought any changes in our favor. We found a narrow strip of open water all along in front of the glacier through which we might pass, but as the river turned at right angles, and ran away from the glacier, we would still have four miles of ice between here and the open water below.

"The whole river, for four miles down, was blocked with solid ice." Photo courtesy of the Wulff Collection, Valdez Museum and Historical Archives.

However, we decided to get by the glacier if possible. This was attended with great danger, for we would often hear a crash near us, as a great body of ice would drop from above, into the river, and should one drop near enough we might be capsized by its waves.

Returning to the boat we ate our breakfast of beans,—which had been cooked the night before, and set out down the stream. Where the current of the river ran, the water was open for fifty yards down into the ice; there we must pull to one side to clear the ice jam.

My boat was in the lead, but as we began to pull for the shore we were caught in a whirlpool. and began to spin around like a top. At the same time we were being borne swiftly toward this opening in the ice, which seemed like a great mouth, yawning to suck us in. As the current swept into this place, great cakes of ice, weighing hundreds of tons, would move slowly around, and coming together, would grind into atoms the smaller cakes of ice which happened to be between them. We felt that if we went into such a place as this, we must surely share the fate of the smaller cakes of ice, and be ground as we had seen and heard them being ground.

We tried to hold the boats, but it was no use; so I said to the men, "When the bow is toward the shore pull hard." By this means, and using almost superhuman efforts, we succeeded in rounding the corner of the ice jam, and thus escaped perhaps as great a danger as we had passed in all our months of daily perils. Not to have succeeded must have almost certainly sealed our doom.

Looking up the stream, we saw our companions in the other boat simply resting on their oars, seeming to be so absorbed in watching us as to forget their own peril, and were surely drifting into the same whirlpool which had caught us.

We motioned to them frantically to pull toward the shore, and they all at once seemed to arouse to a sense of the situation, and began pulling with all their might, but it was too late; they were in the whirlpool.

Immediately they became excited, and kept constantly pulling, while the boat was turning round and round, and in this manner were drifting toward the ice. We saw that nothing could prevent their going into it, and said, "They are lost."

When they saw that they were surely going into it, they began calling to us for help, but we were powerless to render them any assistance. Two large blocks of ice were coming together, and we thought just in time to grind them and their boat, but it passed through and out from between them as they came together with an ominous grind, which sent shudders all through us. Then the great cakes of ice came down against the stern of their boat, forcing it out of the water upon some lesser cakes of ice, and there they were high and dry upon—not land, but ice, and safe, at least for the present.

They noticed that the ice was moving, and that they were being slowly carried around in a great circle, which, if it continued, would in a few hours bring them out near the spot where they had entered the whirlpool, so they sat and waited, and in two and one half hours were in open water again; and pulling across above the whirlpool, they soon rejoined us.

None of these men would like to again pass through the experience of these few hours. The helplessness of their situation was in the last degree pitiful. Sitting face to face with an awful peril, they could do absolutely nothing but wait, whether it was to be ground between the great millstones of ice, or sucked under this four-mile gorge, or swung out into safety; minutes seemed long, yet none too long, for they might be measuring off just the few that remained to them of mortal life.

But we couldn't stop to think of experiences past; new dangers were confronting us at every step of our way.

We began to pass through the narrow strip of open water in front of the glacier. On our right was the great ice jam, and upon our left was this immense mountain of ice, in places as white as snow, and in others as blue as the sky above us.

Here and there were great masses of ice, hundreds of feet in height, which had become broken off the glacier and stood leaning out toward the water, like an immense leaning tower, ready at any favorable moment to splash into the river. While passing these, our boys seemed almost intuitively to dip their oars with a careful stroke, as if an extra ripple might cause the tottering towers to tumble just as we were under them.

What a relief we all experienced when we had passed the last of them, and saw the glacier in our rear. True, we had a field of solid ice four miles through ahead of us, but we couldn't help congratulating ourselves that we had passed this in safety.

It was afternoon when we reached the end of the open water. Going on shore we secured our boats, and ate a lunch of cold beans,—left over from breakfast,—then set out down the river, to investigate, and ascertain the best way out of our difficulty.

The men who had stopped us above this ice jam told us that they had a large boat below the jam, but upon the opposite side of the river, which we might take, if we could get over and get it.

But this was an impossibility, for the river at this point was two hundred yards wide, the current swift, and the wind blowing a gale down the river, and there was no boat nearer than our own, between which and the open water was the four miles of ice jam. And to bring even one of these across these four miles of rough ice seemed an almost impossible task. The afternoon passed without anything being accomplished.

Some of the men suggested that every man abandon all his

goods but fifty pounds, and start on foot with that down the river. This was not thought advisable, for between us and the coast there were a score of streams which could not be forded; and besides, many had valuable skins and other goods which they could ill afford to leave behind.

That night another boat containing five men came down the river, and camped near us. They had less goods and a lighter boat than we, so they decided to portage both their goods and boat over to open water.

They were also short of provisions, and we could obtain none of them. That day and the next we spent in getting our goods across the ice. By that time the men of the smaller boat had their goods and boat nearly across, and we decided to wait until it was clear over, and then borrow it, and cross and get the large one which we had been told was there; this was done, but imagine our surprise and disappointment at finding that it was gone.

After returning with our borrowed boat, we held a consultation, the result of which was that there was but one way out for us; we must bring our own boats, heavy as they were, over this long stretch of ice.

Had the surface been smooth, as ice usually is in the States, this would not have been a very severe task, but this ice was composed—as already intimated—of great blocks of ice, piled up in all conceivable shapes, so that walking over it necessitated about as much climbing as walking. So the reader will readily see what a task was before us. The wind continued to sweep down the valley, bringing with it great clouds of dust which at times nearly choked us.

One day, when we had taken our boat some distance, and had sat down to rest, we suddenly heard a rumbling sound, like the roar of distant thunder. It seemed to come from the center of the great glacier, back many miles from shore, and continued about three minutes. While it lasted, the open water along the glacier's front would rise and fall fully three feet. Whether this was caused by some convulsion of nature, or by the moving of the glacier itself, we never knew.

At noon of the third day we had the pleasure of seeing both our boats again in the open water below the ice. We ate our dinner of beans, loaded our stuff in, and were again on our way down stream.

Only half a mile below we came to Child's glacier. For three miles down alongside of its icy walls the current runs at the rate of about fifteen miles to the hour. This was faster than we cared to go, seated in a row boat, but we were glad to know that our speed only hastened us to where we could obtain some supplies besides beans. We always liked these, but never before were driven to subsist upon them entirely for so long.

Our pleasure, however, was of short duration, for we had barely passed the glacier, when we came to where the river was again frozen over. With ax and oars we broke the ice for a mile, but when night came we had made but a short distance.

Finding a spot on shore well sheltered from the biting wind, we landed, and made a big camp fire, and discussed our adventures of this eventful trip around its cheerful warmth. Another day and our beans would be gone. Then we would have nothing. This was not a pleasant outlook. Still we hoped to be able to reach Alganak before another night. When bedtime came, we lay down upon the frozen gravel and tried to sleep but it was a poor excuse for a night's rest. The wind howled around us in fitful gusts, and the roaring sound it caused, as it swept along the mountain sides, could have been heard many miles.

Our breakfast was, this time, beans and sand, the wind having brought us our seasoning. Then we started again down the stream, having to stop many times during the day to break the ice, which in many places reached from shore to shore. In the afternoon, about three o'clock, we came in sight of a few log cabins, nestled close to the foot of a small mountain some three miles distant. This was Alganak. We all set up a shout of joy, for it was indeed a welcome sight. Think we were childish, eh? Well, it may be we were, but before you pass judgment on our childish demonstrations, try and place yourselves in just the situation we were before we came in sight of this place. We had absolutely reached our last meal of any sort, and had eaten only beans for several days.

CHAPTER XXX

SECURING FOOD

HERE was a prospect of food. Being anxious to reach it, we put all our strength on the oars, but some of our boys were so occupied with planning what they would have for supper, that they forgot to pull. Alganak is situated on a side channel, about a mile from the main stream. We came opposite, and not knowing where to leave the main channel we made our boats fast to some bushes, and started across on foot toward the cabins.

When within a few hundred yards, we came to a stream which we could not ford, and began shooting with our revolvers, to attract attention. Soon we saw a man approaching in a boat; it was the storekeeper. Two trips of the boat, and we were all landed in front of the store.

Near by the store was a log cabin, out of which a family had moved, leaving behind a table and a stove. We were given the use of this cabin while we remained in Alganak. Here the two men at whose tent we stopped near the rapids, left us.

Our party, then consisting of twelve men, went to the store for provision, but intending as we did to stay there one day, we only bought enough—as we supposed—to last through that one day, paid our bill, which was twenty-four dollars, and took our goods to the cabin, and went at it. Loaves of bread and cans of roast beef disappeared like magic, until I doubt if one little kitten could have made a good lunch of what remained.

After supper we returned and brought down our boats. We were now at tide water. The balance of our journey—we were now forty miles from our destination—must be made when the conditions of the tide were just right. So it was thought best to engage an Indian to guide us.

Making our desires known to the storekeeper, he directed us to one who he said was reliable, and we engaged his services for ten dollars, to pilot us as far as Eyak, which was thirty-five miles.

He could not speak our language, but through the storekeeper as interpreter, he told us to be ready to start by one o'clock in the morning.

This was unwelcome news to us, for we had hoped to get a good night's rest here in our cabin; however, we were anxious to get through as soon as possible, so we bought provisions for two days, and at the time appointed were on hand, ready to go.

But when we got there, our Indian friend informed us that the tide had been running down for an hour, and that we must now wait for the next full tide, which would not be till noon.

It was fortunate for our Indian guide that he could not understand the English language, for he would probably have heard things not at all complimentary to himself, for the boys were not in the best of humor at being cheated out of a night's rest which we all felt the need of sorely, only to find that it was to no purpose. However, we lay down and got a few hours' rest. Next time we were ourselves on watch, and just before the turning of the tide, we loaded our boats and started out.

Ten miles down was a fish house, which in the summer was used as a canning factory for salmon, but at this season of the year was vacant. We were to reach this place and again wait for a full tide. Following the course of the stream, we crossed broad flats and wild meadow land many miles in extent.

On the way we saw thousands of wild geese, and countless numbers of ducks, but they kept a safe distance away from our boats and guns, so none were bagged.

About four o'clock we reached the fish house, built a fire, and prepared our meal—dinner and supper in one. From this point to the next fish house at the mouth of Eyak River is twelve miles. This is across flats which are only covered with water at high tide, so at eight o'clock, when it was a little over half tide, our guide came and motioned us to get ready to start. The night was dark and foggy, and how our guide could lay his course, when for five and one-half hours we were out of sight of land, was a mystery to us. But he did and at half-past one we came in sight of the fish house.

Here we were to remain until eight o'clock when the incoming tide would enable us to run up the river. We saw some boats here, and knew we were not alone. Going to the fish house, we lighted a candle, and found five men asleep on the floor. They were on their way out to the States, and were waiting the next incoming tide.

We were soon asleep beside them, but were early up, and ready to start up with the tide when it should come.

The morning was cold and frosty. As we pulled along up the river, ducks by the thousand would rise out of the water ahead of us, and circling around, would light down again in our rear. It seemed to us that this would be a hunter's paradise. We spent little time however upon them, and we now had plenty of provision, besides we were anxious to reach the coast as soon as possible.

About noon we arrived at Eyak Lake. This little lake is a beautiful body of water, one mile wide and four miles long, with mountains rising up from its shores thousands of feet high, and on either side, the base of which is heavily timbered with spruce. Here we were destined to meet another delay, for the little lake was almost entirely frozen over.

Going upon shore we built a fire and prepared our dinner, after which we set out up the lake. Coming to the ice we began to break it and force our boats through. All the afternoon we broke ice, but when night came, were still three miles from Eyak. We must abandon our boats here and carry our goods around the shore to the little village. The shore being a mass of brush and fallen timber, it was almost impossible to get through, so we unloaded our goods, and went into camp for the night. It now began to snow, and soon the earth was wrapped in a mantle of white.

CHAPTER XXXI

SICK ON THE WAY

I HAD not been feeling well the entire day, and at supper time was unable to eat any, so spreading out my sleeping bag on the snow I retired. I had chills and fever all night, and in the morning was so ill that I could scarcely stand without assistance. But there was no possible way of getting to Eyak except to walk, and with one of my companions on either side, we set out. Half carrying me as they did I could go but a little distance without resting, and it was afternoon when we reached Eyak. One of our company had gone on ahead to make arrangements for my comfort, so I had only to go to bed. Here I was able to get some medicine, and by the next morning was considerably better.

Possibly the reader may be able to form some faint conception of the pleasure I felt when for the first time in many months, I took off my clothes, like a Christian, and got into a real bed, and with a real roof over me. Sleeping bags are the best devices known for the trail in that frozen or inclement country, and are perhaps the only safe substitute for a bed on the tramp; but getting into one of them with all your clothes on, except perhaps your boots, and sometimes your coat, is different from getting into a comfortable bed. And the contrast is even greater when one is sick. Just imagine, if you can, how it would seem to you, when almost too ill to raise your head, to be obliged to crawl into such a bed, out on a snow tank or great cake of ice, under the twinkling stars, or the threatening clouds, or in a driving storm of rain or snow. And then the added pleasure of crawling out into a zero temperature several times during the night, if such were necessary. These are some of the "contingencies" of life in such a country. Fortunately they were not of frequent occurrence in our trip, for most of our company were well during the entire time.

I think I can truthfully say that while my first night in a bed was not in any sense in an ideal bed, yet never in all my life did a bed seem to me such a luxury as this one did. To be able to undress, and get between a pair of clean sheets, with real pillows under my head, instead of my coat and boots, one or both, as was true of the sleeping bag, was such a contrast that it was like being almost in another world.

We were now five miles from Orca, where we were to take the boat for Seattle; but we were here told that it would be fifteen days before a boat would be in, as the last mail steamer had sailed the day before.

The boys found a large vacant log cabin, containing a stove and table, and secured it, by rental, for the time we were to wait. They prepared a comfortable bed for me, and I was soon resting in our own hired house.

CHAPTER XXXII

TACKING BOAT FOR SEATTLE

I SHALL never forget the kindness of these men during the few days of my sickness. Had they been my own brothers I could not have received more considerate care. They bought some provisions, and we began to keep house on a grand scale.

Prince William Sound is only separated from Eyak Lake by about half a mile. Over this are built tram ways, upon which handcars are used to carry salmon from the lake, where it is brought on boats, over the sound, from which place it is shipped to all the world.

The first afternoon in our new quarters the boys shot several ducks. They had borrowed some boats, and planned for a grand duck hunt the next day. They were up early, and after an early breakfast, had set out upon their hunt, but had gone only a short distance when they heard the hoarse whistle of a steamboat at Orca. Of course the duck hunt was quickly declared off; and they hastened back to the cabin and began to make the liveliest kind of preparations for getting out of Eyak for Orca.

We had already engaged a man to take us over when a boat should arrive, so in an hour we bade adieu to Eyak and our little cabin, leaving behind all our provisions and cooking utensils, knowing that we should have no further need of them after reaching the steamer.

I was given a comfortable seat in the boat, with nothing to do but watch the thousands of ducks which were flying about and over us.

When half way over, we met a man coming out from the boat after us, the captain having heard that some miners were there waiting to go on the next boat.

We were soon at the wharf, and our goods were transferred to the steamer, and we soon had the satisfaction of stepping on to the fine steamship *Rival*, which was to bring us back to the States.

This vessel was under government charter, and had been to the Yukon River with supplies for the soldiers, and on its return had stopped at Cook's Inlet, Valdez, and Orca, bringing out some soldiers, and a few passengers who were too late for the last regular boat.

Within a few minutes after going on board, she left the dock and steamed out a few miles into the sound and anchored for the night. At early dawn we were on our way out of Prince William Sound, and taking the outside course, the snow-capped peaks of Alaska soon disappeared in the distance.

We will not try to describe the mingled emotions with which we watched them disappear. These mountains, and rivers, and glaciers, and lakes, the hardships and perils, the pleasures and pains, the hopes, anxieties, ambitions, and disappointments which had been crowded so thickly into the past months, were all among the things that were past, but not all among the things to be left behind.

Many, aye, the most of them, were to come back with us to the States as very lively memories, to go with us all the future years, to sometimes be lived over in reminiscences and story, or in dreams of the night.

The invigorating sea breezes soon set me at rights again, and I took my usual part in things, both at the table and elsewhere.

The voyage over to Seattle was an uneventful one; and was in striking contrast to the one we made ten months before over these same waters in our little schooner *Moonlight*.

Our food was the very best, and our sleeping apartments first class. The officers and crew did all in their power to make our voyage pleasant.

CHAPTER XXXIII

EXPERIENCES ON SHORE—PARTING

JUST after dark of the seventh day out, we came in sight of the lights of the city of Seattle, and in another hour we stepped out upon the same wharf we had left ten months before.

During the first part of our stay in Alaska we had allowed our hair and beard to grow, as a protection against the severe cold and the fierce winds and storms to which we were constantly exposed; and when the weather became warmer, we had allowed the hair and beard to remain, to protect us in some degree from the villainous attacks of the ever villainous mosquitoes, who followed us more relentlessly than our own shadows; and when the frosts of approaching winter had delivered us of these, the certainty of what awaited us as long as we were in Alaska made us loath to part with either, so both were allowed to remain until we got back to civilization again. But once here, our appearance made us altogether too conspicuous; for every man of us was almost a veritable Rip Van Winkle to look at. Not wishing to visit a hotel until we had changed our woolly appearance, we returned to the vessel and waited until morning.

Early in the morning I went on shore, and visited first a clothing house, where I purchased an entire outfit; next I visited a barber shop, and exchanged my luxurious beard and hair for a clean shave and short cut and a good bath, paying a difference of seventy-five cents. Then donning my new clothes, I ventured to look at my new self before a good-sized mirror; and the transformation was so great that it was almost past belief, and for a while it was altogether uncertain whether I was myself or some other fellow.

Going out on the streets, I sauntered about for some time, seeing nothing of my companions. Presently a bright, good-looking young fellow touched me on the arm, saying "Beg your pardon, but I am a stranger in your city, and would like to inquire if you can direct me to the Government Steamboat Office." "No, sir," I said, "I can not, for I, like yourself, am a stranger in this city."

He was dressed in the height of fashion, but he seemed more talkative than a stranger would naturally be, and I set him down as a confidence man, wondering what sort of a game he was planning to spring on me. By this time we had arrived at the corner where I turned off on a street leading toward the steamboat landing, and was interested to find that my talkative friend also turned and accompanied me.

I became more convinced than ever that he be longed to a gang of crooks, and determined to watch him. I glanced sidewise at him, to be sure that I would recognize him should I ever chance to meet him again; at the same time I noticed that he was casting sly glances at me, and what was more, was biting his lips to keep from laughing outright. He was none other than George Winters, one of our company, and we had been together during most of the perils of descending the Copper River, shooting its fearful rapids, chopping our way with ax and oars through ice, and dragging our boats over the four miles of solid ice, and all that. But he had exchanged his leather suit, which he had worn all the way since coming to us, for a suit of broadcloth, his Klondike brogans for patent leathers, his slouch hat for a Dunlap, and his Alaska shirt for spotless linen, and his hands were encased in a neat-fitting pair of kid gloves. But the greatest change of all was in the loss of his hair and whiskers. Now he was closely trimmed and clean shaven.

During the afternoon as we came across our companions one by one, we noticed what a wonderful change had been made in each, as he had emerged from the hands of the tailor and the barber; and that night as we gathered at the hotel, few would have recognized us as the same company which had landed from the steamer only the night before.

The next day the time came for bidding each other good-by, and it was with feelings of genuine regret that we shook each other by the hand, and separated to go to our respective homes.

APPENDIX

Listing of supplies Purchased in Seattle for Trip to Alaska, February, 1898

14000	lbs. of	Flour	300	" "		Pearl Barley
1050	" "	Rolled oats	500	" "		Split Peas
1750	" "	Rice	600	" "		Coffee
5000	" "	Beans	200	" "		Tea
1500	" "	Dry salt pork	400	" "		Butter
4500	" "	Bacon	125	" "		Cheese
875	" "	Dried beef	600	" "		Black Figs
100	" "	Soda	750	" "		Pitted Plums
1000	" "	Salt	300	" "		Seeded Raisins
20	" "	Pepper	7	" "		Nutmegs
35	" "	Mustard	350	" "		Soup vegetables
15	" "	Ginger				

1050	" "	Yellow corn meal
1050	" "	White corn meal
875	" "	Granulated sugar
210	" "	Royal Baking Powder
700	" "	Evaporated apples
1000	" "	Evaporated peaches
1000	" "	Evaporated apricots
600	" "	Evaporated onions
2000	" "	Evaporated potatoes
400	" "	Evaporated corn
700	" "	Smoking Tobacco
700	" "	Chewing Tobacco
1700		Saccharine Tablets
600	Pounds of	Candles
20	Cases of	Condensed Milk
400	Bars of	Laundry Soap
100	Bars of	Grandpa Soap
12	Large cans of	Matches
108	Cans of	Extract Beef
4	Large boxes	Macaroni

Total cost of above items: $3,345.54.

INDEX